The R.A.M.S. Library of Alchemy

Volume 38

Sanguis Naturae

Christopher Grummet

Also includes

Universal and Particular Processes

by John de Monte Snyder

R.A.M.S. Publishing Company

Sanguis Naturae

Christopher Grummet

Also includes

Universal and Particular Processes

by John de Monte Snyder

Produced by

Restorers of Alchemical Manuscripts Society

R.A.M.S. Publishing Company

R.A.M.S. Publishing Company
117 Rutherford Lane
Stuarts Draft VA 24477

Sanguis Naturae

First Edition 2015

ISBN-13 **978-1511969208**
ISBN-10 **1511969202**

Image Processing by Philip N. Wheeler

Printed in the United States of America

Table of Contents

Dedicated to Hans W. Nintzel,

American Alchemist

and

Founder of the

Restorers of Alchemical Manuscripts Society

(R.A.M.S.)

Disclaimer

Liability: The publisher does not warrant or assume any legal liability or responsibility for the accuracy, completeness, or usefulness of any information, apparatus, product, or process disclosed. The publisher makes no representation as to the accuracy or completeness of the contents of this book and specifically disclaims any implied warranty of merchantability or fitness for a particular purpose. No warranty may be created or extended by written sales materials or sales representatives. You should obtain professional consultation where appropriate. The publisher shall not be liable for any loss of profit or other commercial or personal damages, including but not limited to special, incidental, consequential, or other damages.

Introduction

Philip N. Wheeler

Christopher Grummet wrote this work in the late 17th century. Subtitled, "Manifest Declaration of the Sanguine and Solar Concealed Liquor of Nature," it was selected by Hans W. Nintzel for inclusion in the R.A.M.S. Library.

There are a few footnotes in this volume that are attributed to "D.H." Although I have seen footnotes under these same initials in other R.A.M.S. works, the person's identity remains unknown.

This Volume of The R.A.M.S. Library of Alchemy also includes "Universal and Particular Processes" by John de Monte Snyder. A much longer work by Monte Snyder is available as The R.A.M.S. Library of Alchemy Volume 31, "The Metamorphosis of the Planets." The symbols in this short work may be more easily understood using R.A.M.S. Volume 21, "Alchemical Symbols."

sanguis naturæ

(Christopher Grummet)

or

A Manifest Declaration
of the Sanguine and
Solar Congealed Liquor
of Nature

LONDON
1696

PRINTED FOR

A.R. and sold by T. Sowle

SANGUIS NATURAE

or

A MANIFEST DECLARATION

of the

Sanguine and Solar

CONGEALED LIQUOR

of

NATURE

by: ANONYMOUS

LONDON

Printed for A.R. and sold by T. Sowle in
White-Hart-Court-in-Grace-Church-Street

1696

TO THE READER

COURTEOUS READER:

Upon my certain Knowledge that the Author (who was a German, & died in that Country, and by whose Death these Two, and a Third Part, came thereto my Hands; which if these Two Parts are well accepted, will hereafter be published) was a True Master of the Secret he writeth of, as well as a Man of great Probity & Piety, & of various Polite & Useful Knowledge & Learning, permit me to speak my thoughts so freely, as to say, That if this Tract doth not relish with you, the Fault is either in your Pallet, that is pleased only with some particular sorts of Meats which you are accustomed to, or else in that your Constitution is such that you cannot bear Strong Meats, & not the Dish set before you.

Book I. Chapter I.

Whosoever attempteth the Search of our Glorious Stone, he ought in the first place, To Implore the Assistance of the All-powerful Jehova, at the Throne of his Mercy, who is the True and Sole Author of all Mysteries of Nature; The Monarch of Heaven & Earth, the King of Kings, Omnipotent, most True & Most Wise; who not only maketh manifest (in the Macrocosme) the truth of every Science to Worthy Philosophers, & liberally bestoweth both Natural & Divine Knowledge on the Deserving & Faithful; but also layeth open his Treasures of Health & Riches (which are locked up in the Abyss of Nature) to those who Devoutly Worship him. And forasmuch as none are permitted to touch the Mysteries of Nature with foul Fingers; therefore it behoveth all who attempt such matters, to lay aside their Natural Blinders (from which, by the Light of the Holy Scripture & a steadfast Faith, they may be freed) that being the means by which the Holy Spirit doth clearly make manifest the most profoundly hidden Light of Nature; which Light alone lays open the way to the Wisdom of Nature, & to unlock the most abstruse Mysteries Thereof.

Chapter II.

All the Masters of Alchemy, who have ever Treated of
this Celebrated Stone, and left us anything in
writing have declared the Matter & Subject (which is
the Chief part of this Art) so obscurely, that
APOLLO himself would be tired in unwinding the
AENIGMA'S they have excogitated concerning it. And
this doubtful Declaration of the Matter is the
reason why many who seek this Science without the
Light of Nature, are precipitated into very great
Errors; because they know not the true Subject of
this Art, but busie themselves about other things
altogether unfit for the Work. But they ought to
consider what the philosophers Stone is in its own
Nature, and what qualities of Their Matters with the
qualities of the Stone, the Thing itself will
discover what is Truth & what not.

1.   The Stone in its Perfection is permanent in the
Fire, and despiseth the most extreme violence of the
Flames.

2.   It containeth in itself, in great abundance the
Vital Fire, & the Virtues & Powers of the Superiors
and Inferiors concentrated in it.

3.   It is resolvable in any Liquor.

4.   It abounds with fixed and Tinging Spirits,
which before its Compleat Perfection were Volatile.

5.   Before its Perfection it hath two distinct
parts, one volatile, the other fixt.

6.   It is of most easie fusion.

7.   It containeth the three Principles of Nature in the highest purity, namely Salt, Sulphur and Mercury.

8.   It containeth in POTENTIA Gold and Silver.

9.   It is made of One Thing.

Seeing the Stone hath the qualities above mentioned, it is plain and evident that the Subject of it ought to have the like-Namely:

1.   That the Subject of the Stone be only One Thing.

2.   That it have in it, in potentia, Gold and Silver.

3.   That it contain in it the Three Principles of Nature.

4.   That it be of most easie fusion.

5.   That it consist of volatile and fixt parts.

6.   That it abound with Tinctures both red and white.

7.   That it be resolvable in any Liquor.

8.   That it be the place of residence of the vital fire, and the Virtues of the Superiors and Inferiors.

9.   That it endure the utmost force of the Flames.

Now let the Seekers compare the qualities of their Subjects with the forementioned qualities, and then they will see whether they are right or wrong.

I know there are many who will not approve of this Description of our Subject; especially Those who are wholly employed about ANTIMONY, VITRIOL, vulgar MERCURY, the perfect Metals, Marchasites, Vegetables, Animals, Stones, and other like things, all which are by no means comparable to our Subject, part of which things are either partly or wholly volatile, or wholly combustible and inseparable by any means or by any Liquor, unless perhaps they are resolvable by a Corrosive. All Philosophers do declare that the knowledge of this Divine Science consisteth in the Knowledge of the Elements and their occult Operations; which is a certain Truth, and it were to be wished that those, who employ their Thoughts about the above mentioned principles, would study This Saying, and follow its direction; There would not then be so many Sophisters, and so few Philosophers; and they would do well to seek out one of the Ancient Philosophers who expoundeth the Elements, and their occult Operations. But this is scarce found in any one, or if perhaps it is to be found, yet by reason of the obscure Stile of the Author, it meets with incredulous Disciples.

And therefore for the Sake of some good men, who perhaps bear an honest mind, I will Discourse something of the Elements, and their operations, and first & chiefly of the mover of the Elements, and of its Life; which not being known, the whole operation of the Elements is unknown.

This Mover of the Elements which, not without good
reason, I will call the living Fire, is Two-fold,
the one Volatile, the other Fixed, residing in the
Center of the Earth, of which at present I will not
speak, but of the Volatile; which is a substance
Invisible, Spiritual, and wholly Fiery, an Eternal
Light nearest to God, the Life of the Elements, from
which the Sun and Moon, the radiant Stars, &
whatsoever giveth forth a Luster in the Heavens,
Takes its Original and Splendor, flying through the
Universe, everywhere present, and most of all in
those Things which stand in need of continual
Nourishment, endowed with innumerable Virtues. This
Celestial Light is Originally most pure in itself,
as long it is not defiled by impure bodies, the
Knowledge whereof is the Sea of Wisdom, which all
who have obtained Light from the Holy Spirit, and
Faith from the Father of Lights, ought to keep safe,
if thy desire a happy Success in this Mysterious
Philosophy. This Light descendeth daily into the
Elements, which are Bodies internally Spiritual,
very simple, and most powerful, which contain in
Them-sieves a certain Seminal Spirit, which is the
very Element; and which Spirit of every Element is
stirred up to motion by the living Fire; and if it
were absent, the Elements would be dead, especially
the Fire, if it were deprived of this fiery
Splendor, which by itself, and not by accident is
the true Principle of Motion in all Things; and to
this the passive Elements are obedient. But yet this
Agent cannot act without the Elements, nor the
Elements upon one another without it. For this cause
the Elements were made, by the most High Creator,
which together with their Body contain a certain
Seminal Spirit, very powerful, which lieth hid as a

Soul in them, out of which by the Action of the living Fire upon it, daily new seeds are produced.

This living Fire, with which the Heavens and all things are filled by the Creator, descendeth through the Elements into the Subject, which is called the Balsom of Nature, ELECTRUM IMMATURUM, MAGNESIA, the GREEN DRAGON, AZOTH VITREUS, the Fire of Nature, the Universal Seed, the Salt of the Earth, out of which all Bodies which consist of the Elements are produced by Nature; and out of this Matter, by the administration of an ingenious artist, by means of a spagyric destruction, new forms of Natural Bodies may be produced; which is one of the greatest ARCANA of Secret Philosophy. For in this Subject lye secretly hidden all the Virtues, properties, & Splendours of Animals, Vegetables, and Minerals, Metals and Precious Stones; which by help of Vulcan, are brought from Darkness to Light.

Now I will describe the action of our living Fire upon the Elements, which descendeth out of the Fire into the Heavens as an Element of Fire, and there whatsoever is lucid or glistering, as the Sun, Moon & Stars, doth secretly derive its Original from this living Fire, & constituteth this principal Element, and obeyeth it as a Son the Father, and a Patient its Agent. And from this Living fire, the Heaven hath its chief power of acting; and is of so great consequence, that if its Action upon the Heaven should cease for one moment of time, whole Nature would be ruined. For the Sun, Moon and Stars would lose their active and influential Virtue, the Elements would not move, and nothing for ever would have any Action; which would be a great mischief to the Earth, and extremely hurtful to all Mixts. For the Power and Virtue of this Living Fire is so great, that if it were absent, the Elements would be dead, especially the Heaven, an' Element which most of all stands in need of this Light. Having passed through the Heaven, it comes into the Air, that great work this Element, and insinuateth itself most intimately into it. In this Element the Virtue of this Fire doth chiefly manifest itself; because in it is inspissated, and constituteth the vital Air; which Air is thus agreeable to the Creatures, for sustaining of Life. For this living Fire simply is not convenient for the Creatures, nor yet the Simple Air; but Fire consealed with the most pure part of the Air, and Air impregnated with the Coelestial living Fire; and so they Constitute vital Airs, which every living Creature receiveth for the Conservation of its life. This living Fire needeth the Soul of the Elements, chiefly of the Air, which it makes use of for a VEHICLE, that thereby it may more easily enter into the other Elements, that is to say first the Water, a subtile and this Element, in which it is yet more inspissated, and taketh a more gross Body of which it standeth in need for irrorating of all Terrestrial Things, especially Salts, Minerals and Stones; all which need such irroeation; (?) thus being clothed with a thick garment, it passeth into the Earth, a dark & thick Element, and of a very powerful fixing Virtue; and there it puts on a saline Body, which predominates over all things, and contains the rest of the Principles, which it had received in the Air, Heaven & Water, that is to say, SULPHUR and MERCURY, by virtue of which it becometh capable of Production. This Salt is the Soul of the Earth, and all other Things.

Therefore if the Earth were deprived of this Salt, it would want the Power of sprouting and budding, which consisteth only in this Salt; of which also Moses was not ignorant, Saying (in the second Chapter of GENESIS) AND THERE WENT UP A MIST FROM THE MIDDLE OF THE EARTH, AND WATERED THE SUPERFICIES OF THE GROUND; which vapour can be nothing else but the subtile parts of this Salt, which hath settled itself in the Centre of the Earth, and by the warmth of the Corporeal Water are made to Ascend, and so do Water the whole Superficies.

Thus we clearly and evidently see, that this living Fire putteth on no other Body than that of Salt; because that alone is fit for generation, and so the Balsome of Nature and generated by the action of the Fire upon the Elements.

Now I will speak but a very little of the fixed living Fire, which is hidden in the Earth or Center of the World, and there hath taken up its most fixed Habitation; and by many Philosophers is called the Corporeal Water; but it may better be called the Fire of Bodies. To know this is the most Secret Mystery in all our Philosophy. This fixt Fire hath a great Sympathy with the volatile Fire; for it wanteth it as an aliment, and to its Nourishment, which is continually attracteth out of the Water and Air, and converteth into its own Substance; and in this as in a Center all the Virtue lyeth concentrated, which being scattered, flyeth in the Circumference; as may be observed in Man, in whom this Fire fixed in the Center of the Heart, hath its seat as the Yolk in the Egg. But its operation is invisible and very Secret, and yet very powerful, which also few know; for it operates by its heat in all things, which lie in the Earth, and exciteth the Flux and reflux of the Sea, as the. Pulse in Man is excited by the Fire which lieth hid in the Center of his Heart. Hence also all the Watery and Airy Vapours, by the help of this Fire are elevated from the Earth and Sea into the Air, which compose the Clouds, and by rare-faction of the Winds (being impregnated by the Vital Spirit) fall down again to the Earth in form of water. And thus every Searcher of Nature must acknowledge it to be true, that there is only one Subject under the concave of the Moon, in which alone the Virtues as well of the Superiors as of the Inferiors, lie concentrated; out of which by the Chemical Art, Stupendious things may be brought to pass.

This Body is Salt, but not common Salt, or any other Salt of this kind, but a SATURNINE & MINERAL SALT, which hath chosen to itself a residence in the Sphere of SATURN, and is also called the Heart of Saturn; out of which being made clean and bright, and purified from all Excrements, by an easie Art, a certain gummy Liquor is drawn, called by the Name of Glorious MERCURY. But you must be wary in the choice of this Salt, there is only one Salt which is useful to us, a pontick fiery, bitter and Mineral Salt of a SATURNINE Nature, out of which this famous Liquor is extracted; which is of so great moment, that without it, no Transmutation of Metals can be made. In this MERCURIAL Liquor may be seen what is SULPHUR and MERCURY; for the Sulphur at first sheweth itself in a Purple and Yellow Colour; but the MERCURY is invested with a watery and Airy humidity; and tho' the Salt appeareth not, yet its Virtue is eminent in this Liquor.

For it is wholly Saline, and an easie Fire is
coagulated into a permanent Earth, which
representeth Salt.

And so you have Three Principles of Nature: Salt,
Sulphur and Mercury.

These are but a few things which I have said of the
Matter; which tho' it be very secret, yet its
Operation of it is more secret which nevertheless in
my following Discourse I will reveal, so that its
occult may be made manifest only to Men Elected by
God.

Chapter III.

I have said above, that the living Fire (with which the Heaven and the Elements are filled by the Creator) doth secretly descend into the Subject which is the Balsam of Nature, and the absolute Subject of the Philosophers Stone; whose nine properties I have declared, and which I think fit a little to unfold; in the search of which the Ancient Philosophers were very long employed. This Natural Body in which Sol and Lime do inhabit, they found; as Sufficiently appeareth by the Writings which they have left us; out of which the Modern Rout of Alchemists who seek the Golden Stone, have conceived as many CHIMERAS as their Brains could reach; whereof some affirm that the Concretes of the Vegetable and Animal Kingdom, others that Minerals as ANTIMONY, SULPHUR, & MARCHASITES, and the rest of the Minerals; others, that Metals themselves, Gold, and Silver; but others of a more Subtile wit, .that VITRIOL & common Salt is the Subject of the Glorious Stone; which opinions the Sincere Searchers of Nature ought to leave free to their Authors; and let him know by the Light of Nature and Experience, that the chiefest Error of these Smatterers in Chymistry is, that they falsely persuade themselves, that there are divers Subjects of this Art, contrary to the several Rules of the Philosophers, who teach clearly, that there is only one Subject, and say that this is compounded of the Four Elements, out of the three Principles of Nature; and so both the four Elements, and the three principles of Nature compose their Stone, by which they delude these foolish Chemical Novices. For they say, the Stone is made of one Thing, of Two, of Three, and of Four: By which Contradictions they declare to the Sons of Art, and

Pious Searcher this only wonderful Subject of this Art. For immediately this Thing which Composeth the Stone is but one; which is divided into a fixt and a Volatile, into an Agent and a Patient, and so it is Two; and although it be so divided, yet it doth by no means lost its Unity.

So also when it is divided into Salt, Sulphur, and Mercury, and so is three-fold, neither doth this Division destroy its Unity; much less doth the Division into four Elements do so. And tho' this matter be common, yet it is not known to these Novices in Chemistry, who seek it among Animals, Vegetables, and partly Minerals but they know not that in these it is already determined; among the Minerals, the chief are ANTIMONY, Vulgar MERCURY, and VITRIOL. In ANTIMONY indeed there is found a MERCURY, but too Combustible; (read Coagulated) and also a SULPHUR, but too Combustible. In VITRIOL there is also a Mercurial Nature, but too Acid, and hath no incombustible, Sulphureous Salt. In MERCURY and other Minerals there is found a useless proportion both of SULPHUR and MERCURY, of which the greatest part is in part wholly Volatile, or wholly Combustible, and so is not fit for this Art, but our only MINERA, which is inclosed in an Earthy Substance, out of which may be plentifully drawn the Philosophers AQUA PERMANEUS, whose Virtues are also powerful as well in Art as in Medicine, in no sort Venemous, but it is the chief Purger of Humane Bodies, by Urine and Sweat, the highest Medicine for the Venerial Disease, Cancers, Leprosy, Fistula's, and other incurable Diseases. And its Virtue is great in a QUARTANE, the Dropsy, Stone, Gout; it most powerfully resisteth all Poisons and Philters.

But of what kind this Matter is, which is endowed with such Virtue, and out of which is plentifully drawn the Famous Philosophical MERCURY, I have sufficiently demonstrated in the preceeding Chapter, and herein will farther demonstrate; which is not properly Mineral nor Vegetable or Animal; yet a Metalline SULPHUR, SALT, & MERCURY, are together purely and plentifully in it, which is obvious to everyone, and lieth hidden in everything, but especially in the Earth, the receptacle of all the influences, in which also the Virtues of the Sun, Moon, and Stars are found corporally. This the Artist ought to take where it is most near and most pure, in form of a Triune Salt, which elsewhere I called the Salt of SATURN: Out of this Salt groweth Gold, and all other things, in the whole Mineral Kingdom, into it they may be resolved again. And as in Gold lieth hidden a bright and diaphanous SULPHUR, so also in our SATURNINE MINERA, there is a coelestial SOLAR Fiery, Diaphanous red and sweet SULPHUR. For where there is glittering Brightness, there is light; where is light, there is Heat; where is Heat there is Life and very powerful Action; and which is a great matter, in it reign its Elements animated with a living Fire, which is a Coelestial vivifying, Fertil and greening Spirit; the Light, Force and Life of all things.

And although the Coelestial Sun doth much help the production of Sublunary Creatures, yet without this Internal Sun nothing can be generated; which also the philosophers knew. Therefore all other Concretes are rejected, as also Salts, except one which is the SALT of WISDOM, Power and Strength, and the Mother of the other SALTS, namely our CENTRAL SATURNINE SALT, a SULPHUREOUS and MERCURIAL SALT; whose Heart

26

and Blood doth easily dissolve every metal, and coagulate MERCURY.

The Fire as Sol and Luna, tho' they appear not to sight, yet they are powerfully in the inward parts of our Matter, and it possesseth the Seminal Virtue of all Things; so also that unless what is hidden in it be made manifest, they will not appear; which is done only by reduction & purification of the matter, that the FECES (which beclouded our SOL & LUNA) may be throughly purged away, and the matter may first grow white as a Diamond, & be as fulgent as a Ruby, then they appear to sight. Which reduction must be made with a certain contrary Liquor; for Sol & Luna, which are secretly in our Matter, & rule powerfully in it, are not reduced so as to appear to Sight, unless this reduction is made by a contrary, which is a Menstruum or Most Subtile Vapour, penetrating and resolving, containing in it Air, Fire, & Water, & separating the pure from the impure, & yet first extracted out of our Minera; which Liquor possesseth only power of reducing, manifesting, & multiplying Tinctures; & therefore it is called The Secret Fire of Nature, which alone exciteth & perfecteth Tinctures. But yet we must not therefore suppose that the Subject is red or any other colour, but it is white in appearance, & red only in Potentia; because this Nature of redness lieth secretly hidden in the Belly of the Air of our Matter, not shewing its Tincture visibly, because if it be put into the Fire, it cannot manifest a fixed Tincture, unless it be first reduced by an ingenious Artist, so that the Watery & Earthy heterogeneous Substances may be separated; then appeareth a Tincture resisting the Fire, & shewing itself in it White & Red.

The certainty of this Solar Subject may be evidently known, if out of it the three principles of Nature can be separated. What they are I will explain.

SULPHUR residing in our Matter, is its fiery, most subtile, and most thin part, partaking of a subtile Earth, which indeed is the perfect & absolute Tincture, having power of rubifying and illuminating every Body, by reason of its innate oyly, very fat, unctuous and viscous, fiery and ethereal quality; upon which account it is able by its subtile and internal action to produce all Natural Things. Which SULPHUR is called the Philosophers Secret Fire, the living Fire, in the luminous part etc. Therefore if anyone in the Anatomy of our Matter see a certain shining, subtile, clear Substance, full of a fiery shining redness like a Ruby, and full of active Virtue, he may be certain that he hath seen our SULPHUR, and the Secret Fire of the Philosophers. MERCURY is the Aereal and unctuous humidity of our Matter, and the inseparable Companion of SULPHUR, and is as a Menstruum to it, cherishing and nourishing it, and a MEDIUM of conjoyning the SULPHUR with the SALT; but the reason why it is Viscous as SULPHUR, is, because it hath subtile Earthy parts resolved in it, which it took with it in the Anatomy made by external Fire. The Salt is the Principle of coagulation; and coagulateth the MERCURY and the SULPHUR, and in which a new form is introduced by the action of the SULPHUR which operateth in it, which SULPHUR is very bitter and acid, in which bitterness there is a certain fiery Substance corrupting the inward parts of the Salt, and which being corrupted immediately it receiveth a new form, and that a living one, which is a great Secret.

These Principles are also very much defiled with
heterogeneous Feces; which an Artist ought to know.
SULPHUR in the first place, aboundeth with
destructive and consuming Feces; But the MERCURY
with watery and cold substances contrary to life;
and in Salt there are caustick, viscous and bitter
Salts; all which must be separated and if not, they
occasion damage and unlucky Success to the Work.
This one Minera is of easie fusion, so as it can
catch MERCURY upon the fire before his flight; which
if it be circulated by the Philosophers Wheel, so
that those parts which hinder speedy fusion be
separated, and the Elements firmly caogulated, it
becomes of more easie fusion than it was before;
which fusion dependeth upon the Saline and
Sulphureous Spirituality, which is a perfect and
concentrated Light, penetrating every Body and on
all Sides illuminating it with the Tinging Rays with
which it abounds. The Spagyrical Art by Distillation
and Separation, affords us Two Sulphurs out of one
Minera of a Saturnine Nature; one is volatile and
green, the other fixt and fiery, which two by
circulating their Elements were made one, which
Sulphurs are of the Nature of Salt produced by
Nature out of the most pure Soul of the Elements.
Because by the mixture of the living Fire and this
Soul, these Sulphurs are produced in the greater
World, out of which by the Chemical Art the two
Sulphurs of the Philosophers are extracted, which
consist of the most Subtile and pure part of the
Sulphurs produced by Nature. Out of this pure
Substance Metals are also generated, which differ
only according to the purity of the place, and the
more or less fit disposition. Out of these Two
Minerals is plentifully extracted the Mercury of the

Philosophers, which is their Radical Humidity mixt with a Subtile Earth. For as by this Earthy Sulphur is meant the heat and fire of Nature, as also the forms of the Matter, which we also call Sulphur. So also the Humidity of this Substance is our SULPHUR, in which, (if they be joined by Art in a certain proportion, and are decocted in a glass Vessel, circulating their Elements; first Water, then Air, Fire and Earth, and then they are purified by reduction into a certain chaotick, thick and viscous Mass; then by Distillation into Liquor, one white, another red and shining as fire; lastly they are fixed into a glorious and permanent Earth) consisteth all our Art. SULPHUR is the principal part of our Tincture, and that which plentifully beareth rule in our Matter, is two-fold, as we have declared, white and red, fixed and volatile. The fixt is the GREEN LYON; which lieth hid in the centre of our Concrete, abounding with fixt and tinging Tinctures, but the volatile is the Fire of Nature and our SULPHUR, full of Power and Efficacy of Tinging and illuminating, as this testment doth manifestly declare. For it is the blood of our GREEN DRAGON, distilled from the very Bowels of it, abounding with redness; therefore it is not without cause, called the blood of Nature, which stirreth up its own SULPHUR, lying hid in an Earthy Substance, and brings it from Power into Act; and then out of it do arise our two glorious MERCURIES, our two perfect SULPHURS, one red, the Husband; the other White, the Wife; Springing out of one Minera. And that Sulphur which we call the GREEN LYON, is the Fire of Nature, which lieth hid in the Center of our Subject, understand Salt, and there is detained shut up in a strong Earthy Prison, disabled to exert its force, unless by its associate it be set at liberty

30

from its Fetters, so that it may come out together with his Companion. This deliverance consisteth in Solution, which is very difficult, for this SULPHUR. which we also call the Stone, is both most hard and most soft in its Nature, and therefore it is not easily dissolved, except in its own Liquor, that is his Companion in which it is most soft; so that it can be set at liberty only by this aery Companion, which otherwise could not be delivered, neither by Fire nor Water; which is a Secret known to few, of which I will speak more particularly hereafter.

This fixt SULPHUR is very powerful, and sustaineth everything that operateth in Nature, but as soon as it is set free, it ceaseth from its labour, if together with its Companion it be carried aloft, and in the Top of the Vessel, where if they are detained, they constitute a certain Substance bright as LUNE, called DIANA; at this time I say, it receiveth the power of Transmuting. Since the Stone is of the Substance of Salt, it resolveth itself in any Liquor; but the Salt, out of which it is prepared is of most profound research, and differeth much from other Salts; for it is fluid in the fire, and values not its Martyrdom at all, and loseth nothing of its Virtue therein, though it be kept in it divers years, which other Salts, as VITRIOL, SAL GEMME, and other the like Salts, cannot do; for by often repeating ignitions, they all turn to an unprofitable and unfluxible Calx, which is not resolvable in any Liquor, being of the Nature of damned Earth. Tho' the Matter to sight is most vile and most cold, yet its more inward parts are meer Fire, and abound with the Living Fire, and the Virtues both of the Superiors and Inferiors, and therefore its Soul flyeth in all places to bring down the Living Fire. For the Father of it is the SUN, and the Mother the Moon, from which it secretly deriveth the Virtues of all things. This Living Fire ruleth powerfully in our two Radical SULPHURS, fixed and volatile; which two being firmly united, do constitute our Universal MERCURY, which containeth in itself the two Central Fires of the MACROCOSM, the Coelestial and Terrestrial; and these Two, by the help of External Fire, are reduced into one Substance, in which the Coelestial and Terrestrial Virtues lie concentrated, which heat is the innate heat of everything, which often sheweth its splendour in the Eyes of Fishes, Hairs of Brutes and Men, in Insects generated out of Dew, as also in rotten Wood; but I think it is sufficient, that everyone daily seeth the luster of Gold and Silver, Pearls and Precious Stones, and also beholdeth the SUN and MOON. Lastly, it is to be considered, that the Tincture of the Philosophers, is a substance Tinging imperfect Metals in a very strong Fire, into perfect Gold and Silver, from whence it necessarily followeth, that its Subject ought also constantly to resist the Flames, and to rejoyce in them. But it is not the whole Substance of the First Matter which endures the Fire; because it aboundeth with many Elementary Feces, which are combustible in the Fire; but only its pure parts, which are also called incombustible oyls, rejoyce in the Fire, and are permanent in it: because they are of a pure Nature, and not defiled with any Feces, therefore the Fire cannot touch them. Wherefore it is necessary in the first place, to purifie the matter, and take away the Sphere of SATURN, which becloudeth the SUN and Moon, before they can despise the Fire; and then decoct these parts till they be reduced into one Thing, whose Virtues neither Fire, nor Water, nor

Iron, nor Air, can diminish, but they, unvanquished, resist all their force. Consider therefore, O Man, the Wisdom and Power, which the most Wise, Eternal, and Omnipotent God, JEHOVA, hath granted thee. Consider these things in Humility of Heart, and Sing Hallelujahs to him without ceasing, for, HOLY, HOLY, HOLY, IS THE LORD ZEBOATH; THE HEAVENS AND THE EARTH ARE FULL OF THE MAJESTY OF HIS GLORY, AMEN.

Chapter IV.

We have spoken copiously of divers things necessary
to this Art; but we will speak yet more, and things
more necessary, among which is the Solution of Gold
into Water, which is the beginning of making our
Divine MERCURY, and that is, to convert the hard and
soft Nature of our Gold, into a thin and Watery
Substance, with Conservation of the Internal Nature,
and Property of Gold. For if this internal SULPHUR
should be corrupted and destroyed, it would not be a
Physical, but Sophistical Operation; but that which
we desire to perform, ought to be done with our
corrosive pontick bitter and sharp MERCURY, which
imitateth the Natures of SOL and LUNA, with which We
truly dissolve;, because our SULPHUR is hard and
kept close in the Prison of its Excrements;
therefore this Substance ought to set it at liberty
and extract it, in the meantime by the same
operation we also calcine, reduce, dissolve, and
putrifie the Gold. For if we calcine, the Fire
burneth the Heterogeneous fetid and viscid parts
which naturally adhere to Gold, and conserveth the
Homogeneal parts, full of life, and so attenuateth
them, that thereby they become of more easie
Solution. But nothing doth more destroy and conserve
these parts, than our pontick and corrosive MERCURY,
by reason of its fiery Nature wherewith it abounds,
which also the Philosophers call the Fire of Nature;
and if they speak of Fire, they mean this Water, and
not common ravenous culinary Fire, as appeareth in
their Books, when they say, BURN OUR COPPER WITH
VERY HOT FIRE. AZOTH and FIRE are sufficient to burn
Laton, and yet this burning is done with a gentle
Fire, for with a very strong one it would be
destroyed, because our MERCURY being tender and full

of FIRE, which would make it fly away from our hard and unattenuated Gold, and the Gold would remain undissolved, and if it were dissolved; it would ascend with its MERCURY in form of a red ponderous Water.

Therefore also in this Operation two Works are performed; namely, a coagulation of Mercury, or of the fiery Substance into a red Mineral and viscous Water, and a Solution and conversion of Gold into the same Water; which the Philosophers hint when they say, that Mercury cannot be coagulated unless the SULPHUR be dissolved; and on the other side, the SULPHUR cannot be dissolved unless the MERCURY be coagulated; they must mutually act one upon another for the performing of this Operation (for Gold, which is hard and strongly compacted, needeth this Subtile and Aereal Mercury) which if it be accomplished, out of it is generated, by putrefaction, a middle water, powerful in Tinctures, which is that permanent Water which the Philosophers so earnestly desire; that is to say, that with which, and the glorious Salt, is generated a certain Substance which we call DIANA regenerated, and the triumphing SULPHUR of Nature.

And it is to be noted, that this regenerated DIANA is generated of the fiery Salt and fiery Water, leaving behind it unprofitable Ashes, and is so fiery and penetrant, that it can burn Gold; and without it no ELIXIR is made. For in our glorious Salt there is a certain Virtue which is fiery, subtile, penetrant, and most mighty in Power, which is found, in its last and utmost destruction in which is the Virtue and power of multiplying Gold: and it is so powerful, that this Virtue can neither be burned by the vehemence of a calcining fire, nor be extinguished by the coldness of the dissolving and washing Water, nor be stirred by any force of the Winds; And therefore, saith a certain philosopher in these words:

"Our Stone is alone ponderous, unmoved by Fire, unmoved by Water, unmoved by wind; and it is also most light, hollow and spongeous, moved by Fire, moved by water, moved by Wind, because it is moved and altered by the Spirit which is called Fire, which is called Air, which is called Wind."

Our Gold is not vulgar Gold, which is sold by goldsmiths, or anything like it, but it is a certain other substance more precious than Gold itself, whose green and golden Colour doth sufficiently demonstrate its original and Excellence. This Green Gold in its first Root is clothed with a foul Garment, which must be separated by dissolving it by help of Mercury of Gold, first extracted out of Gold, and abounding with a bright golden Sulphur; which alone is capable of performing this Solution; because it dissolveth nothing but the golden nature of Gold, which is of its own Nature. But the Earth adhering to Gold is not of a golden nature, and therefore is not dissolved by this Solvent, but falleth to the bottom of the circulatory Vessel, in form of a muddy and viscous matter, very like to the Sediment of Urine, and is easily separated from the dissolved Gold. But this GOLDEN MERCURY is wonderfully intricate to be searched out, and tho' it be found everywhere, yet it is most difficult to be found, by those who know it not, tho' easie to those who know it, and know its Nature exactly. For it is a white and serene, ponderous, Acid and pontick Liquor, of an ethereal Substance, which is sublimed with a most gentle fire, and converted into Air, and there in a glass Vessel turneth into Water, which is that so celebrated Animal, Vegetable, and Mineral MERCURY, which the Ancient Philosophers knew, especially LULLY, who doth not without cause, call it Red Wine, and LUNARY. For it doth not alone constitute the Essence of these three Kingdoms, but also the Heaven itself; and all the Heavenly and Earthly Natures have their Being and Conservation from this Spirit and Watery Substance, by reason of its living fire with which it abounds, and without which no Creature can live.

Therefore the Ancients call it a fiery vigour, the natural living fire, Animal, Vegetable, and Mineral; by which all things in the Air, Water and Earth, which have life, are nourished, and which failing they die; by which Substance is performed the True solution of our Gold, and by no other; because all other Waters are Heterogeneous to our Gold, and therefore not fit for our Work. Of this kind of Gold we have spoken. Now I will also say something of the Body, which is a certain, Saline, Sulphureous, fixt and permanent Matter, which must be dissolved with a Philosophic Menstruum, else all the pains is lost, which are taken in any Operations, and they are all Vain who boast of such a Tincture, without this Menstrous and permanent Liquor, which alone is able for penetration, Subtilisation and depuration of this Body, and reducing it to the highest Purity. For this Water was before naturally included in this Body; which if it be again poured upon it, it opens its Pores, and attenuateth without any inward hurt; because it is of the same nature with the Body, so that it can do nothing else but nourish and conserve it. This Water is extracted out of the very Bowels of our Matter; for even our Body was Water before, and by Nature, by means of the internal fire, is reduced into a Body, which is easily reduced and resolved by its own Water, whose Nature it before had, had, which if they are dissolved and decocted by fire of the first degree, convert themselves into a viscous fiery penetrant Substance, which by farther operation passeth into an Earth, which by the Philosophus is called the black Earth of MAGNESIA, whose Operation they have hidden with the utmost envy. Our humid Mercury, which containeth the fire of the Elements, is extracted out of our only Minera, by force of external fire, which being highly purified, is poured again upon the Body and conjoined with it so purified, and worketh lying upon it, until at last it exalteth it to a higher degree; namely, into LUNA of LUNA, which being again dissolved and putrified with our Water, affords us a permanent Water, which resolveth all Metallick Bodies yea and Precious Stones; with which Water, and our glorious Earth, we prepare our Two SULPHURS & TINCTURES both white and red.

But that this operation may be more easily understood by all, imitate the following Manner.

PRAXIS.

Take our Corporeal Mercury, Animal, Vegetable and Mineral, pure and purged by the fire from all Excrements, and put it into a glass Vessel, stopping it very firmly, and digest it by assation, till the bloody Pores of this Body be opened, then take it dry and dissolve it in AQUA FOETIDA, white and ponderous, which is also Vegetable: digest it for some time till it appear a dry mass. Dissolve this again and filter it well, decoct it till the whole be resolved into a bloody Liquor. Shining and ponderous; circulate this Wheeling about and extracting it into a liquid, hard and this Substance which is our Mercury, with which Gold and Silver are burnt into Ashes. A great and wonderful Mystery, known only to ADEPTS; the Secret Knowledge of which Earthly MERCURY, Hermes hath shewn in his Table, Saying, "His Father is the Sun, and his Mother the Moon, the Wind carrieth it in its Belly, the Earth is its Nurse; it ascendeth from Earth to Heaven, and again descendeth into the Earth, and receiveth the Virtues of the Superiors and Inferiors, its Virtue is entire, if it be turned into Earth."

Out of which Earth which is our Mercury Sublimed, glorious and fixed, is made our Elixir.

Now let us return to the Water, by means of which this our Subtilization is made; which is a certain Water very subtile and precious, acid, foetid, corrosive and sharp, which the Ancients hid under the Name of Vinegar, as also of other acid and fiery Liquors, as of Aqua fortis, Vitriol, allum, Saltpetre, and Sal Armoniack; by which Water our Body is made Subtile, and reduced into the next matter of the Stone, which is a Viscous and muddy Water, fiery and full of Tinctures, with which the Stone sublimed, Viscid and fixed into Earth is fed, that it may ascend to a Royal Dignity. Which Secret, namely of subtilising our Body, the Ancient Philosophers would not reveal, but left it to God to reveal it to whom he pleased; nevertheless they left in writing among the Rubbish, some hints how that middle Substance is to be prepared, yet very obscurely (namely that Spiritual Substance which they named with many Names) and yet is the Key and Foundation of this Subtilization, of which Water it is said in TURBA, Our Body must be environed with the Flames of our red fume and be broken by it, as being a fire against Nature; for by this Water, which is full of Fire, Our Body is washed till it be also made a Mineral or permanent Water.

But that I may give you an Epitome of this Chapter, I say, That the whole Work of our Subtilization consisteth in Vapour and Water, which is called a Whitening and purifying water; which I divide into two parts, namely the water of the two ZAYBETHS, white and red, whereof one calcineth the Body, and in Calcination coagulateth Itself with it; but the other purifieth it from its blackness, whiteneth and rubifieth, and at last makes it volatile. Which water is called Acetum Acerrimum, because it is very sharp and acid. This Humidity also containeth in itself an unchangeable Tincture, which can by no means be extinguished. This Water is called AQUA VITAE, VEGETABLE, ANIMAL, SPIRIT OF WINE, Strong VINEGAR, SATURNAL WATER, and many other Names.

But the Artist, who endeavours to set upon this work, must know that every Body is dissolved by a sharp Spirit, and made volatile with a Spirit; and if the Spirit be so prepared by the help of the Body, our MERCURY is prepared, which putrifieth, washeth, and fixeth and incereth itself, till at last it attaineth to the highest Subtilty and purity, and sublimeth itself from the bottom of the Vessle into a white Stone. This must be separated from its Feces, by sublimation and reduction; and then will be prepared the foliated Earth more white than Snow, which after its due Decoction, coagulateth and fixeth vulgar MERCURY, and transmuteth every imperfect Body into true LUNA. This most precious Whiteness is our Arsenick, an incomparable Treasure, which above all other things the Philosopher needeth. This SULPHUR must be calcined, till it be converted into a dry and very subtile Powder; which Powder must be imbibed with the white Oil of the Philosophers divers times, till at length it floweth like wax and then there will be prepared the White Stone, whereof one part Tingeth a Thousand parts of any Metal, into true Silver.

Thus it appeareth clearly, That all that can be desired in Philosophy, may be extracted out of only the Body, and its own Sulphureous Mercury, which two with the help of the fire, accomplish all things; and he who understandeth these two, understandeth all that is necessary to our Art. Tho' the Philosophers say many things of divers matters, yet they mean nothing else but these two Sulphurs, which for the sake of the Sons of Art, I will explain.

Chapter V.

Among the Secrets of Alchemy, the greatest is to draw Water out of a Rock; Verily a hard and very difficult Work, unless CHYMISTRY alone had shewed us the possibility of this thing; which the Artist ought to endeavour to do by Fire, which in the beginning must be gentle, in the middle strong, and in the end most vehement; so that all the Aereal and Ethereal Spirits of this Rocky Minera, may issue forth into a fit Philosophical Vessel, and there resolve themselves into water; which water with wonderful Sympathy loveth the Rock, from whence it issued; which Water is called by various Names, as ROCK-WATER, ARGENT VIVE, A FUME, THE TINGING COELESTIAL SPIRIT, INCOMBUSTIBLE SULPHUR, WINE VINEGAR, SUCCUS ACACIA, SPIRIT OF WINE, TEMPERATE WATER, THE LUCIFEROUS VIRGIN: all which Names signifie this water; which if it be again conjoined with it, remaineth Stone, and often operateth resting upon it, it acquireth a wonderful active Power, as all know who are acquainted with this Water. This operation is also called by the Philosophers, a destruction of the Compound; which destruction is not to destroy as the Vulgar Chemists think, who destroy Mixts by Corrosives, but by the unlocking the Bonds of our Compound, by which it is bound, which if they be unlocked, it is divided into parts with conservation of those parts which constituted this Elementary Mixture; which parts so divided; are purified and delivered from Excrements and Impurities, with which they abound in their Composition.

But that this might be more easily done, the Ancient Philosophers devised this Distillation and

Destruction, by help of which, the parts might be most highly purified and exalted to such a degree of Purity, that there upon a new Compound might be made, of greater Efficacy. But to bring this to pass, the Artist ought to follow Nature, as all Philosophers, both Ancient and Modern teach, and to extract our Mineral out of the Bosom of Nature, where she hath hidden it, and purifie it most subtily, by very frequent Cohobations and Reductions. For so it throughly sheds all its Excrements, and whatsoever else hindereth it from its perfect Power of Transmutation, which is wonderful, and yet it is more wonderful, that in this vile and abject Minera, lieth hidden the celebrated Stone of the Philosophers, whose Essence also by reason of its obscurity nobody can see, unless it be delivered Therefrom, and brought to light; for before it is set at liberty by the Chymic Art, it is a rude, vile, abject, and undigested Mass, which is also found scattered in the Earth, out of a hundred pounds whereof, scarce one or two pounds can be extracted, which is the pure Soul, Fire, Oil, and powerful Tincture; so also but one pound of our glorious substance; which, after many Martydoms, we extract out of our Minera, and after every Extraction dissolve, coagulate and fix; till passing through almost all other colours, it appeareth white, subtile, dry and penetrant; which abovementioned colours do sufficiently evidence the Essence of this Minera, whereof the chief colour is green as a most certain indication of life. These two Substances, that is to say, the Mercury and glorious Earth, are sufficient to perfect the Stone, having first, as we have said, accomplished their purifications; because our Sol and Luna before that, were involved in obscurity.

The Excrements of the fixt Body are Earth and Fire,
burning, viscous, insoluble by our Mercury, and
therefore they are easily separated in our Water,
and those things which have the nature of the Body,
do easily mix invisibly with the water, all those
things appearing which have not the Nature of the
Body, which puddle the water and confuse it, and
which by a quiet rest of the Vessel, fall to the
bottom, and there unite, and are separated from the
limpid water, which retaineth this precious Body in
its Bosom, which at length, by reduction, appeareth
again, and by Assation, is more and more attenuated;
by attenuation is more and more cleansed from its
Earthy and Viscous Excrements, which as before, are
separated by our Water, till at last there remaineth
a spongeous, fixed, most pure Body, But this Water
is a thin and Viscid Water, abounding also with
Excrements, which do naturally adhere to it; for
these are a fiery and Sulphureous Earth, able indeed
to coagulate this our Water in a long time, but yet
of no moment, which nevertheless many have unluckily
magnified, the Philosophers exclaming, In Mercury is
whatsoever the Philosophers Seek; which is not to be
understood of this Water, but of our glorious
MERCURY, which notwithstanding is extracted out of
this Water, which containeth Fire dissolved in it,
for which reason it hath Power of Coagulating
itself, which is a long Work; therefore to quicken
the Work we dissolve some parts of the glorious
Earth in our MERCURY, that the Secret may be
compleated in a shorter time. But this thin and
Viscous Substance which we also call our Mercury,
doth also abound with many Aereal and Watery
Excrements, which savour of the nature of Fountain
Water; but there are others, which are of a greasy,

oyly, and fat nature, and are the corroding,, and caustic Fires of a sulphureous nature, which must be separated first, by a gentle digestion, in a Vessel exactly well stopped, that thereby they may better be let loose, than by Distillation and Filtration till no cuticle at all swim upon the Top of the Water, which may easily be seen; for that would be hurtful to the Water, and bring damage to the Work; but if this MERCURY be rightly prepared, it is fit to perfect the Mystery which ought to be accomplished, and to perform many other Operations; but chiefly to perfect the Sublimation, which cannot be done without pure Materials. For the Body admitteth not unclean Waters, and Water agreeth not with an unclean Body; therefore both ought to be clean, that they may be perfectly united and at last Sublimed to the top of the Vessel, and there constitute the SULPHUR of NATURE so much desired. This operation, MORIEN showed in these words, "If you do not perfectly cleanse the unclean Body, and do not dry it, nor whiten it well, and do not mix its Soul with it, and do not take from it all its ill scent, till after its cleansing, the Tincture cometh into it, thou hast discovered nothing at all of this Mystery."

Therefore we must apply ourselves with our utmost endeavours to this Purification and Mixtion, that both may be united and joined pure together with an inseparable bond; and a durable Matrimony, which even the Fire may not be able to Separate.

Chapter VI.

We have said many things of Purification, Solution
and Distillation and that we may proceed father to
things necessary to this Work, we will speak
something of Philosophical Calcination, which among
the Philosophers hath been of great account. For it
purifieth those things which before were involved in
the darkness of Excrements, and it bringest to light
clean things, which before were stained, and
affordeth to us oyly SULPHURS profitable only to our
Work; but not as the Vulgar Sophists do, who attempt
to Calcine by violent Fires, AQUA FORTIS,
Cementation, and the like which are plainly contrary
to our Calcination, who yields dry and Calxes not
flowing like wax. This is not our Calcination, but
rather the loss of our Body; because they do not
increase but diminish the innate Fire of our GOLD,
which alone we want for perfecting the Tinctures.
They who Calcine thus, are blind, and walk in
darkness, for our Calcination is not a dry
dessiccating of our Body, by which the Body is made
dry and not flowing. This is not our Calcination,
but after we have drawn out all the Stinking
Menstrous Spirits from the Mineral Body, and
abstract and cohobate till at last we obtain a Body
pure, fixt, fiery, and fluid as wax; out of which
(being resolved in our Mercury, and so often
cohobated upon it till it be turned into a red and
viscid Oil) is prepared the permanent Water, and the
glorious shining Earth, the only Pillar of our
Tincture. Thus our Calcination is the augmentation
of the innate fire, and the highest Purification of
the Body; which is done by our Pontic Water full of
fire, which burneth and mortifieth the body, after
Death brings it to an Immortal Life.

Here perhaps the rout of Vulgar Chymists will object, not understanding this Calcination, and for that reason will say, How can Calcination be made with Water, seeing the fire is the only Instrument of Calcination? To this we answer, That Philosophick Calcination is not the Calcination of the Vulgar, which Calcineth Mixt into Ashes, dry and deprived of all the innate fire, and fit for no work necessary to Life, as above said; but our Calcination, Calcineth the Mixt into a viscous Humidity, abounding with fire, and fixt and Permanent in it; which Humidity alone Alchemy useth to perfect her Arcana, this made with the pontick Water, full of Living Fire, which alone is capable of perfecting this Calcination, which defending Bodies from the most violent flames of the fire, and mixeth itself with their internal Fire, and fortifieth it, which hidden Calcination is known to few, and the True Knowledge of it is a great Secret in this Art. Which that the Sincere Searcher may more easily understand, let him take the hard Vegetable Body, in the bottom; and take it fresh and most subtilly Powdered, and put it in a Vessel well secured, and put that into a Furnace with an open Fire, increasing the Fire by degrees, and at last giving a most strong Fire, let him urge it so far, that the Vessel comes plainly to a candent heat, so all the Watery and stinking Vapours left by the Menstrual Spirit, will pass out, and the Body will be freed from them; but the Fire must not be encreased to that degree to make the matter vitrifie by the Flux, for thus it would lose its vegetable Virtue, and the operator would lose his oil, and Labour, and the Body would lose its thirst and hunger of Drinking up its proper Soul. Therefore it must be calcined with

47

very great Caution, and so that it may retain this
Thirst; and thus the Calcination will be rightly
performed; which is a very tedious and long work, in
performing of which the Artist ought to be very
Cautious. Now, after the Earth is so prepared, take
it and work it by help of our Calcination till it be
wholly freed from all its Earthly and burning Feces
by Reduction, Solution, Calcination, and Imbibition,
till by Calcination it becometh wholly red, and the
Calcinatory Water be also freed from all its phlegm
and Watery Humour. Take the Earth now fluid, porous
and plainly fiery, and grind it into Powder in a hot
glass Mortar, grinding it over a Fire of Ashes for
two or three hours, until it be a subtile Powder;
then add to it drop after drop of the Aqua Vitae,
grinding it continually with a glass Pestle till the
coagulating fume of the Earth be pretty well
satiated. Then put it in a glass Vessel, digesting,
and imbibing with Aqua vitae, and grinding till it
be Converted into a bloody, glorious and MERCURIAL
Liquor; which Liquor is AQUA VITAE regenerated by
the Fume of our Earth, the Coelestial Water,
Ethereal Liquor; and this is a Short and Secret way
which few also have known. The other way is longer,
and is thus, Take the Earth rubified by virtue of
the Fire of our Aqua vitae, and grind it subtily,
and digest it with its water, till it be converted
into a black sparkling Mass, which is the ANTIMONY
or BLACK LEAD so much spoken of by the Philosophers,
which is made in three months, then Wheeling it
about, and circulating it well, work it till it
becometh a Tincture Citrine and red. This way is
long, and lasteth almost two years, and is very
tedious, which also the Ancient Philosophers taught,
saying: AZOTH AND FIRE ARE SUFFICIENT; FIRE AND
WATER WASH LATON, PURIFIE, AND FIX AND INCERE IT.

And wash LATON and tear your books, least your Hearts be broken. Which way also a certain Philosopher teacheth darkly, saying: "Take that which is most Volatile, and conjoin and wash the more fixt with the volatile, till the most fixt receiveth the most volatile; then turn the Earth into Water, the Water into Fire, the Fire into Air, and inclose the Fire in the middle of the Water, and the Earth in the Belly of the Air; Mix the hot with the moist, and the dry with the cold, because one Nature overcometh another, and Nature rejoyceth in Nature; and afterwards Nature containeth Nature, but the Earth containeth them all. For when the four Natures have ascended up to Heaven, again at length descended, so that the Fire may descend into Air, Air into Water, Water into Earth; but the end of the whole Work is Powder n' Ashes."

These and the like words the Philosophers use to describe their Secrets, by which they delude the ignorant, and cast a mist before the Eyes of the Vulgar Chymists but as I said before, let all lay aside their Opinion of our Calcination, who believe it is done by the Vulgar way; these Persons are daily deceived, and deceive many with themselves; let them learn first, before they attempt our Calcination, which is of so great moment, and so wonderful, that in it Fire alone and Azoth are sufficient, and know, if they desire to know, that every Spirit is fixt by a Calx of its own kind; which if it be fixed with the Body, it Calcineth it, and if the Artists so Calcine, they will find it profitable; but if not, Sorrow and Sadness will overwhelm them, because being Ignorant, they dare attempt our Calcination.

49

Chapter VII.

Because the Vegetable Body which we also call
Mercury, is of a vile Nature; namely, Earthy and
Watery; therefore it ought to be exalted to a more
noble and subtile Nature, namely Airy and Fiery,
which two are very near Principles of this Mercury,
as well according to the intention of Nature as of
Art, and therefore the vegetable Body must enter
again into the Belly of its Mother, thus by Death
and Regeneration it may attain to such Dignity; but
which cannot be done but by Philosophic Corruption
and Alteration, which causeth our Menstrous, Fiery
and Airy Vapours and Fumes (which before came out by
Distillation from the Body) to thicken by a gentle
Digestion and Rotation, that this Water being
circulated, may the better penetrate the Pores of
our Body, and so successively alter the inward part
of the Body, and at length truly and rightly
regenerate it. This Putrefaction or Alteration of
this Body, consisteth in Solution of the Same Body
in its own Airy and Fiery Vapour, which can best by
Digestion, alter the Body, and bring it to a new
Generation. And it is altered whilst it is dissolved
in that Water, because this Water is the true
Sephicher of the Body, in which it dieth and is
putrified. For this Water, and no other, can alter,
putrifie, dissolve, distil, calcine, and mortifie
the Body, until at length it is reduced into a most
subtile, not Terrestrial, BUT VISCOUS ALCOHOL, which
is done not only by dissolving this vegetable Body
in its own Water, but by many other labours and
operations; NAMELY, BY DISSOLVING IT INTO WATER, AND
THEN AGAIN DRYING, CALCINING AND INHUMATING IT, and
this again drying and calcining and afterwards
distilling, till at length the Body as it were

INVISIBLY by so many, and such operations, is truly
altered; the sign of which is a dark blackness,
which is the true mortification of this Vegetable
Body, in its Mother or Menstrous and vapourous
water; which is done in the beginning of the work,
and in the crude Conjunction of a pure Agent and
Patient; which is a hard Herculean and hazardous
Work, the Knowledge of which dissolveth all other
Arcana of the following Operations; but especially
of the second Alteration, which is done with our
Sublimed and glorious Sulphur; by Inhumation and
Imbibition, in a philosophical Vessel, with our
permanent MERCURY, of which we will not now speak;
but of the first, which is very laborous, and
requireth an Ingenious Artist; of which also the
Ancient Philosophers made no mention at all; which
whosoever understandeth, very easily attaineth all
the rest, in which no Man can err, if after
Distillation and Inhumation, he prepares the Earth
to Citrinity and Viscosity; of which Body so
prepared and calcined to a Citrinity, take one or
two pounds, and Powder it Subtilly in a strong
Morter, and imbibe in the same morter from hour to
hour, grinding it subtilly, and imbibing with our
Living Water, till the Matter be converted into a
fat and slimy Mass; whence you must circulate till
it be thin, and circulate again till it be thick,
sometimes imbibing and distilling. So by reiterated
Works, this Earth will become a thin and viscid
Mass. Take this and put into a glass cucurbit, which
put in Balneo, and there circulate it till it be
turned into blackish Ashes, which you keep carefully
and dry them in a gentle Fire, in a glass Vessel.
Then take these Ashes powdered, and put them into a
glass Vessel very well luted, and distill at first
with a gentle Fire; then somewhat stronger, and so

will ascend our MERCURY white, viscous and limpid,
which we call the LUNARY VIRGIN MILK; now increase
the Fire, and there will ascend a gummy Liquor, red
as blood, and transparent, which is the blood of SOL
and our Earth, which is extracted from part of the
Body and Soul of our Stone. This is that Liquor
permanent and Triumphing over all Metals and Stones,
the blood of the Green Lyon, the Secret Fire, which
must be extracted from its crudity, and exalted with
the glorious Earth; to exalt which, take it pure,
and pour it upon the remainder (which is Lead
already calcined from redness to black) and digest
it upon this Lead, till it hath extracted its Salt,
and be satiated with that Salt; then it must be
exalted, which is our triumphing, exalted and
glorious MERCURY, of an Hermaphroditical Nature,
which is that Water which putrefieth, purifieth,
coagulateth, fixeth, distilleth, calcineth and
incereth itself; which is so Secret among the
Philosophers, without which no Tincture can be made
which an Artist can use to make AURiJM POTABILE. Put
into it Vulgar Gold, having passed the Royal Cement,
and being then most subtilly foliated, circulating
the Gold and distilling, till it be converted into a
thick Oil, splendid as a Ruby, the use of which
reneweth youth, and restoreth debilitated strength.
But for ELIXIR, take it and circulate it upon SAL
ARMONIAC sublimed and fixed into a citrine colour
(N. B. not Vulgar SAL ARMONIACK) and circulate till
it be fixed, then ferment and multiply, until this
MERCURY together with its Earth, flow and remain
fixed in the Fire, Tinging every Metal into True
Gold.

This now what I was willing to say concerning
Alteration, which alone containeth the hidden

Secrets of Philosophy. For our Stone must often die and be revived and regenerated, and at last attain to the highest glory; which we have at present .so clearly laid open, that he must be of a dull wit who doth not perfectly understand it. I have written clearly, and will yet write more clearly; but it will make many admire apprehending that I break the Seal of HERMES. But let those Know, that I have written clearly to the Sons of Art, to whom I would lay open more, if it were Lawful to do it publickly, but to the Mysophilosophists and Sophists, these will be meer Enigma's, in which is more, they will not believe there are so great Secrets hidden in Nature.

Chapter VIII.

For the Exaltation of our Body (which we also call Gold) that of it may be made a new Heaven and a new Earth, it is necessary that the Body being already made pure, be again joined with a pure Soul, that so both being perfectly united, may be exalted and glorified; which glorification, that it may be rightly performed, it is requisite that the Body be made pure by Death and Separation, and that the Soul be likewise purified, to do which, the Artist must in certain quantity (but cautiously) pour the Soul upon the Body, so that the Soul being so joined with the Body, may carry it to Heaven; and so both are perfectly divested of all Excrements, and acquire a very high penetrating Virtue. Both must be freed from Excrements, because this Soul (as is sufficiently shewed already) needeth many purifications, by Subliming it, that it may be freed from all its Original Uncleanness, before they be united, so as they may become one thing by Glorification. For if they are conjoined whilst they are impure, they will never be united, because the Original uncleanness with which they abound, would hinder Union; and their Conjunction being hindered, they could never unite, for in that Union consisteth the Glorification, but both, that is to say the Body and the Soul, are separated from their Original Uncleanness before they are conjoined, not by one and the same method, but by divers; that is to say, the Body by Death and mortification and often reiterated, and the Soul by Sublimation often reiterated; but it must be observed, that the Soul must by little and little be poured upon the Body, and be cherished by natural heat, till the same hath imbibed all the Body, and the Soul which is

contained within the pores of the Blood, recieves a
Sanguine Body, which it doth receive when the Blood,
in which it is contained, hath imbibed the Body, and
so the Soul and Body are united by immediate
Contact, and being of the same Nature, one hath
easie ingress into the other, and then the natural
heat cherisheth them so, that they are more and more
united, and by union become one Body different from
the former, in which the Soul and Body are exalted
together. And it is to be noted, that the Body
before was gross and foul, and the Soul in like
manner impure; both which are now purified and
united, which if they be united, by help of the
Blood in which the Soul is contained, are exalted
into a fiery Body, much different from the former,
which is the Son of the Fire, a glorious SULPHUR,
not unlike to shining TALC, out of which is
immediately made the Physical Stone. Now let the
Searcher of this Science consider, how great a Work
the glorification of the Body with its Soul is; and
let not any one accuse me of obscurity; I say, when
this Soul, known to all, with the BLOOD OF THE BODY,
are truly and really conjoined, then you must take
this Matter, and put it in to a glass vessel well
luted, and a glass Alembic very well closed, and
digest it, then give a good strong Fire, and so our
Sulphur will ascend to the sides of the vessel, and
will leave a black powder in the bottom, very
volatile and of no value, which is the damned Earth
deprived of all that is good for anything. But if
this Powder be heavy, it is an evident sign, that
still there remaineth something good in it, which
could not be dissolved, and then this Body must
again be imbibed with the Soul and the Blood and
again be sublimed, till it ascendeth white as Snow,
and shining, which is our fiery, foliated SULPHUR,

which alone we need, to make any Tincture; to which, for abbreviating the work; we add pure Luna dissolved in the permanent Water; then we decoct, fix, incere and ferment in a close Vessel, Till it be compleatly fixed, pure, flowing and Tinging.

Take the dead and living Body, and put it in a glass Vessel, and pour upon it its Soul till the Body be all imbibed by it, then distill and Sublime; reiterate this Work often with fresh Water or Soul, till the matter sublimeth itself clear as a Star, which you must take and put it into an Egg with a long neck in hot Sand; digest it for a week, the next week increase the Fire, at last increase it more, and so it is fixed; Take this Sublimed and fixed Mercury, and dissolve it in the fire against Nature; cohobate till both become one Water, in which dissolve LUNA, and decoct, imbibe and fix till they flow, because it Tingeth Venus into Luna.

Chapter IX.

Because I have declared to all Lovers of this
Science, the beginning and end of perfecting our
MERCURY, which is the chief and longest part of our
ELIXIR, which being had, all the rest may be easily
performed; therefore I will speak of its Perfection
and Operation into an ELIXIR; whose first operation
to accomplish this end, is thus, Take our Earth very
highly purified, which is our Gold, hollow and
spongy, and put it into a glass Vessel, and there
irrogate it by little and little, with its own
subtile Humidity, which easily entereth this
spongeous Body so that by means of Circulation, the
Airy and Fiery part of this Subtile Humidity may
incorporate, and be coagulated, and be united
together with the Earth. Then irrogate again the
subtil Humidity, and circulate for eight days in a
Vessel very well closed (and here above all things,
beware that you do not irrorate (?) this Earth but
by little and little, from eight days to eight days,
in a very long trituration, so that the force of
this Water may not suffocate the Virtue of the
Earth; because the Virtue of the Earth is weak in
the beginning of the Imbibition, which if it should
be suffocated with abundance of water, it would
become an unprofitable mass, void of all Action).

But the phlegm of our subtill Humidity may be drawn
out by Alembic during the Circulation. So by
reiterate Irrogations and Circulations this Earth
will become a pure Fire and Aether, and the Artist
will then obtain his desire, when the Earth is
rubified by the Spirit and Soul, which be Irrogation
and Circulation, have united themselves together
with the Earth; which if it be distilled, he will

have the BLOOD of the GREEN LYON, the SATURNAL
WATER, which is the first Operation of the MERCURY,
in which the Spiritual Substance is transmuted from
Nature to Nature by means of the Body, so often,
till together with the Body, they constitute the
Permanent Water.

The Second and last Operation, is that of the
FIXATION of the permanent Water and the glorious
Earth, of which the Philosophers say: "That it is a
commixtion of qualities, a Copulation of
Complexions, a Reconjunction of things separated, a
Coagulation of Principles, a Disposition of what is
repugnant; which must be done by a gentle Fire,
Cherishing the parts mixt together, and put into a
glass Vessel, being first made very pure.

And in the luternal Fire of these parts being
excited by a gentle External Fire, doth dissolve and
decoct them, and by decoction they are again by
little and little inspissated and made thicker,
until at length they are wholly fixed, and remain
fixt in the bottom of the Circulatory."

For the Earth containeth in itself a Fiery most
thin, dry and insensible Fume, which coagulateth the
Volatile part, being of its own Nature and
Substance. This Fume lying hid in the Center of the
Earth, by its Action converteth the other volatile
Elements into its own (namely a fixt) Nature; and
then the Motion of these Elements ceaseth, because
they have attained this disired end; which if they
be again dissolved by the volatiles, their motion
beginneth again, till the fixed have overcome the
volatile. Then again motion ceaseth, which if they
are dissolved again, they work afresh, & etc.

Here all operators must observe, that in this
operation a Two-fold Fire must be used, the one
Internal, the other External, which External must
not over power the Internal; the Internal is a dry
Mercurial Etherial NECTER, and our glorious MERCURY,
which vivifieth, conserveth, and nourisheth the
Matter, and bringeth it to perfection: This Fire is
not moved but by our External Agent, which if it be
slow in Operation, the Internal Fire lieth Still,
and produceth nothing; but if the External be too
Strong, either the Vessels break, or the Matter
burneth; therefore the Fire must be warily applied,
so that the Fumes, which lie hid in the Centre of
our Earth, may be moved, and then the Spiritual
Humidity will resolve the Earthly Siccity, and the
Earth will be impregnated by the volatile, and will
grow thick; the Sign whereof is blackness. And if
the Spirits of this compound be more inspissated,
various colours will appear, and by a farther
Operation, there will appear a white colour,
afterwards a citrine, and lastly, a red diaphanous
colour; and after reiterate Operation, the Matter
will be of easie fusion, fixed, and Tinging all
imperfect Metals into pure Gold; which that the
Artist may attain.

Take our glorious shining Earth, and fix it
Philosophically, as above we have declared, and put
it in a fit glass Vessel, let it be dissolved there
in our Water against Nature which also is called
LUNARIA, the BLOOD of the RED LYON, distilled Spirit
of Wine, SATURNAL WATER, our glorious MERCURY;
digest the solution for three weeks, then open the
Vessel, and join to it an Alembic, and distill by
Fire of Balneam, all insipid phlegm that can be

distilled, and when it ceaseth, take away the Alembic, and shut the Vessel; put it again to circulate, then all the Humidity, by little and little will be fixed, and will grow thick like mud of a blackish colour; circulate it farther till perfect blackness appear., and by farther operation whiteness, and lastly the highest shining redness; which is the fiery RUBY, Tinging and healing the Leprous Bodies of Metals; the multiplication whereof an Ingenious Operator can easily effect. Namely, if he dissolve the Stone of the first Order, completely finished, in our glorious Mercury decocteth, fixeth, and incereth; and so he may multiply It, and very highly exalt it; which that they may accomplish, I wish to all, by our LORD JESUS CHRIST.

AMEN

Book II Chapter I.

I have, in the preceding Book, sufficiently taught, not only the Theory, but also the Practice, sufficient for understanding the operation of this Divine Science. But for the more clear understanding it, by divers Praxes as well in general as in Particular; I have written this Second Book for the benefit of the Faithful and Worthy, as also of those who have attained to some Knowledge of our Mysteries, that they may more easily obtain their desire.

And against petty tricking Chymists, who endeavour to make the Tincture of the Philosophers, in one Vessel, for very little charges, in one Furnace in a short time, and shamelessly, and with a brazen Face, fraudulently profess this Art, which they are not in the least worthy to Know. And here I would advise the True and Faithftl Searchers of this Art, that they understand that there is but one thing in Nature of which all things are made, which can be desired in Philosophy; which tho' sometimes I have called and shall call CALX VIVE, sometimes TARTAR, sometimes VENUS and other Names, yet I say, that only one thing is to be understood, as with me beareth witness, the Ancient Philosopher HERMES, saying, As all Things were from one by the mediation of one; so all things proceeded from this one by Adaption.

The First Praxis.

The Masculine Earth of Sol, of itself, can bear no
Fruit, as the Male without the Female cannot have
any off-Spring, and therefore necessarily the Male
needeth the Female; Our Solar Earth needs the Water,
which is its Female. Take therefore, IN THE NAME OF
THE CREATOR OF HEAVEN AND EARTH. This SOLAR GOLDEN
AND RUDDY, and add to it the water of Dew, which is
its Wife and Mother; for this Earth is generated by
the Dew, and put it into a round glass Vessel, so
the Earth will resolve itself in the Dew-Water, and
the Water will be impregnated with the GOLDEN SEED
of the male. Then give a gentle Fire of Circulation,
so the superfluous and stinking sweat will vanish
out of the Vessel, which being gone, if the Female
begin to fly and follow the Sweat, close the glass
firmly and continue the Fire, so the Matter, feeling
the Fire, will work, namely the water upon the
Earth, and by long Operations and Continuation of
the External Fire, the water will extract the Seed
of the Earth and grow thick, and wholly by the
farther continuation together with the Earth, will
thicken into a blood red Liquor, which is the first
Fruit of the Philosophic Tree. Take this and
circulate, evaporating its superfluity, adding more
water and circulating, until the Earth with the
Water be turned into Air and Fire. Then distil,
first the Air, which reserve firmly closed up in a
glass vessel. This Air is the white Air, a vivifying
and unctuous Air, the life of Metals. Secondly,
distill the Fire, which is RED VITAL FIRE, a Fire
vivifying the Souls of Metals, keep also this Fire
apart. Now rectifie first the White Air till it be
bright and serent as Crystal; in like manner
rectifie the Fire till it be like a pure Ruby; Then

take the Earth and separate from it the Water, which
rectifie and join it partly with the Fire, and
partly with the Air. Rectifie the Earth by drying it
gently till it be white. Then add to the Earth,
first, the Fire conjoined with the Water, and
circulate the Fire upon the Earth, till the Earth
appear plainly dry. Add again the Fire with the
Water, and circulate as before till it again be
plainly dry; and if all the Fire be coagulated by
the Thirsty Earth, the Earth will be turned into
Fire. Now add the Air with the Water, and circulate
till the fiery Earth hath swallowed up the Air; add
again Air, and circulate till it be again dry. Then
add all the remaining part of the Air, and circulate
for some days, and take out your watery, fiery and
Airy Earth, and put in another Vessel, and give
gentle Fire and a certain watery humour will arise.
Then put an Alembic upon your Vessel, and distil,
increasing the Fire, and so there will pass over,
first, an airy, fiery and earthy Water Splendent as
Luna.

Then cease, and put to it another Alembic, and
distil an airy, fiery and watery Earth, which two
being had, if the Artist be adapted, he hath enough
for doing farther things, if he proceed after the
following manner. Take the airy, fiery and watery
Earth and pulverize it, grinding it subtilly in a
glass mortar, and put it into a glass Vessel,
imbibing this Earth with the airy, fiery and earthy
Water, grinding this Mass strongly upon a gentle
Fire, till it be like a thin Paste.

Now distil and circulate it till it be thin, and
liquorous; then distil, and pour this distilled
Water again upon what remaineth, and distil and

cohobate until the water with the Earth becometh a fixt Oil, which must be circulated upon Gold after this manner. Take Gold calcined most subtilly with SULPHUR VIVE, into a purple red Calx, which put into a glass, and pour upon it of this thin Oil, and circulate in a close vessel till the Oil become red. Take this and fix it upon the remaining Earth of Gold, after a Philosophical manner, till both are fixed into Powder, which resolve with the Oil aforesaid, and convert it by circulation into a fixt Oil, whereof one part Tingeth much Copper into Gold. This manner is very difficult and long, by reason of the many Purifications and long Circulations and Distillations and Conversions of the Elements, but the following manner is shorter, which is done with Calx Vive.

The Second Praxis with Calx Vive.

Take Calx Viva, calcined to a redness by a strong Fire in a dry Reverberatory; put it into a Vessel, with a strong cover, adding SPIRIT OF WINE, and imbibing with the said Spirit till it will drink no more, then distil the Phlegm from the Spirit of Wine, which being passed over, increase the Fire, and join another Receiver, and distil the SPIRIT OF WINE from the Calx Vive; and when the CALX VIVE is plainly dried, then take it and dry it more in a glass vessel with a good strong Fire, and being again cooled, add SPIRIT OF WINE, and distil; but if you see a skin swimming upon the SPIRIT OF WINE, separate it by filter, because it is a combustible Sulphur. And so cohobate the Spirit of Wine upon the CALX VIVE, always separating the Phlegm, till it be thick, oily and fat, then cease, and take the remaining CALX VIVE, and calcine it in a

Reverberatory with a strong Fire, until it be plainly white; put it so white into a strong glass, and imbibe it with the thick water, reiterating till the water be coagulated by the Fire of the Calx vive; then digest this Mass four days, and distil first a Water which is AQUA VITAE, from RED WINE, SPIRIT OF WINE, rectified, etc. And when that is distilled, increase the Fire, and change the Receiver, and so there will ascend a VOLATILE SALT, WHICH SALT is the Terrestrial Fire of the CALX VIVE, purified, coagulated, and made VOLATILE by the SPIRIT OF WINE in form of a bright Salt, which the operator must take, and assating, calcine and imbibe it with the AQUA VITAE of the Red Wine, and then dissolve that Mass, and distil till both become one Water, shining as Crystal; which is the fiery Mercury of Calx Vive, resolving all Metals.

This Praxis of CALX VIVE is shorter than the former, but in working it, the Artist ought to be Ingenious, especially in Calcining the CALX VIVE, and imbibing it, which must be done warily. And to this PRAXIS the Vitriolization of Tartar is not Inferior, which is not only useful in all Tartareous Diseases, and in resolving their Obstructions, but also in increasing the Anima's of Metals.

The Third PRAXIS, of Vitriolate Tartar.

Take SALT OF TARTAR very well calcined, and well purified by Resolutions and Calcinations till it be Porous; dissolve it by Imbibjtions with SPIRIT OF VITRIOL, then dry and imbibe, and again dry so often, till one part of the TARTAR coagulateth two parts of the SPIRIT OF VITRIOL.

Then take it, and powder it, and spread it upon a glass plate, and set it in a moist place to resolve into an oily Liquor, which evaporate in a glass CUCURBIT in BALNEO till it be like Honey. To this add more of the SPIRIT OF VITRIOL, and dissolve this Honey-like Mass, and when it is dissolved, distil oft the Spirit of Vitriol, which pour again upon what remaineth, cohobating so often till the TARTAR, together with the SPIRIT OF VITRIOL, become one Water; which take (for it is the fiery Water of TARTAR and VITRIOL) and distil gently in Balneo, first the burning SPIRIT OF VITRIOL; then encrease the Fire, and change the Receiver, and distil the OIL OF TARTAR, which must be rectified, as also the burning Spirit of Vitriol; which Two, are our MERCURIES sufficiently fitted for the Composition of the Elixir of the first Order. There is yet another manner very Subtile, which is done by Extraction and Sublimation; but it is very Secret, which I will also Communicate to the Worthy, and it is done out of the VITRIOL of VENUS.

The Fourth PRAXIS of The Vitriol of Venus.

Take Vitriol of Venus which is made of VERDEGREASE and distilled Vinegar, by Extraction, as is known; Powder it, and put it into a glass Retort very well luted, put it into a Furnace with Sand-Fire, and put to it a Receiver, and begin to distil first with a gentle Fire till the Phlegm be come over; then increase the Fire, and when the white fumes begin to distil, change the Receiver and join a new one, which must be well luted; and when the white Spirit is distilled, increase the Fire; and as quick as can be, change the Receiver, and distil the RED OIL, which is the OIL of VERDEGREASE; encrease the Fire till the Retort be of a white heat, and when no more will distil, take off the Receiver, and break the Retort, being first cooled, and take out the CAPUT MORTUUM, which is obscurely red, and ponderous, by reason of the VENUS which it containeth; powder it, and pour upon it its dephlegmated Oil, and also its white Spirit; and when you have poured on all the Spirit, close the Vessel and circulate these Liquors upon the Earth till they are perfectly united; then distill, and first will come over a white and gummy Liquor, which is the exuberate water, then encrease the Fire and there will ascend the SULPHUR of VENUS, subtile and penetrating all Metals after its Calcination. Take this and powder it, and put it to the exuberate water, circulating and dissolving till both are turned into a glutinous Liquor shining like Talc; which circulate till nothing will ascend and descend, then distil, and there will distil a serene Liquor, which is out triumphant and exuberate MERCURY; and when it ceaseth to distil, encrease the Fire, and a white SULPHUR will distil, which is the glorious SULPHUR extracted from the Earth of our VENUS, and the Feces which remain are the TERRA DAMNATA. The following manner is of the Salt of SATURN, useful and very profitable upon Metals, by reason of the grain of Gold which it containeth.

The Fifth PRAXIS of the Salt of Saturn.

Take of the SALT of SATURN very well purified, and mix it with two parts of VITRIOL Calcined. Put this mixture into a CUCURBIT well luted, and join to it an ALEMBIC, luting it strongly, and distil into a good large Receiver first with a gentle Fire, and the Water which distilleth, is called the WATER AND OIL OF NATURE distilleth from the heart of SATURN, which rectifie well till it is bright, break the CUCURBIT, and if the CAPUT MORTUUM is red, it is good, if not Calcine it in a CRUCIBLE with a gentle fire till it be red. Take this and separate all heterogeneous things from it, as well as may be, after the Vulgar manner, till it be pure; which take and put into a large glass Vessel, and pour upon it its distilled Oil in great quantity, and put the Vessel in a warm place for four or five hours, and then filter what is dissolved of the CAPUT MORTUUM, and upon what remaineth, pour new Oil and filter what is dissolved; pour all the Solutions together, and distil all the Oil by Retort, so a certain white and subtile Salt will remain in the bottom. Take this and dissolve it in new Oil, and cohobate this Oil upon the SALT of SATURN, till after the Oil the volatile Salt of SATURN riseth; which purifie by four sublimations, every time changing the Vessel; and taking out that which is pure; and rectifie the Oil by Seven distillations. Then conjoin the Salt with the Oil, and digest this Mass for four weeks in a vapourous Bath, then distil in a Retort well luted, cohobating so often until they are inseparably united. Which Oil is our MERCURY, which being decocted with the Anima of Gold, and fixed, giveth great Tincture upon Lead. But the following manner gives not place to This.

The Sixth Praxis.

Take Urine putrefied, and inspissate it, out of which so inspissated, make a Salt which is an Animal Salt. Distil this in a very strong Retort, and what distilleth rectifie seven times, till it be pure and very bright, which keep.

Take what remaineth in the Retort and calcine it, and extract out of it a Salt with common Water, which salt must be rectified by Calcinations till it is white and floweth. Take this and Powder it very subtilly, and dissolve it in the water reserved as above, and when it is dissolved, abstract all its Superfluities, seal the Vessel and work the matter by Circulation of the Elements of this urine until the matter appear dry by means of Circulation, which dissolve again in the abovesaid WATER, and circulate the Solution by Distillation till all be converted in the bottom of the Vessel into a very thick and fat Oil, which dry and distil, and there will distill a two-fold water. One white, the other yellow, each of which rectifie seven times by itself. Now take the Oil remaining in the Retort from the Distillations of the white and yellow Water, and sublime it in a clean Vessel, increasing the Fire by degrees, and take what is sublimed, and put to it the above-said yellow rectified water, and circulate the water with the sublimate till they are united; in which dissolve Gold, and cohobate the water upon the Gold till the Gold be turned into an Oil, which is augmented by the yellow water, conjoined with the sublimate, in INFINITUM. And it is to be noted that after every cohobation of the water upon the Gold, the Phlegm must be separated; this way is very available to Metals.

The Seventh Praxis.

Take our Vegetable Body, which is our Gold extracted from the MINERA OF SATURN, Powder it very subtilly in a very clean glass, and if it be one pound, put to it of our water (which is the Aqua Vitae distilled from Wine) two pounds; mix it very well with a glass Pestle, grinding it for two or three hours continually, and when the Mass is so mixed, put it in a good strong glass, and digest it for a fortnight: Then open the Vessel, and evaporate the superfluous Water, which is the phlegm of that wonderful Wine from which the phlegm distjlleth first, and when the Mass is dry, powder it and again dissolve it in this Water, and again digest, and again evaporate, and when the Mass is dry, yet once more do with it as before, thus put it into a Cucurbit, and give a Fire of Sublimation and what is sublimed put together, and what remaineth, work over again with our wonderful Wine, and sublime, and what is sublimed, put with the formet, and so often work the remainder with our wine, till there remain in the bottom a subtile Powder of no value, for it is the TERRA DAMNATA, and the Desert laid waste, which cast away; but take that which is sublimed, and sublime it seven times by itself; then Powder it most Subtilly, and put it into a glass, and put upon it so much of our glorious MERCURY that it may become a Paste, which so work that out of it by Circulation, may be made a fixt Oil. This Tingeth all impure Metals into Gold or Silver, according as it is fermented. Our glorious MERCURY is commodiously prepared after the following manner.

The Praxis of the Glorious Mercury.

Take our corporeal Mercury, which is Animal, Vegetable and Mineral, the reason whereof for the present I will not speak of.

Powder this most subtilly, and pour upon it the Water of the Rock in equal weight (for this stony Spirit is white, and contains the Soul of the

70

Elements, and therefore it is called the BLOOD OF
NATURE, Secret, extracted from its own Body, Animal
and Vegetable) and digest it in BALNEO; circulating
this Water upon the Mercury till it will work no
more; then separate the Water from the Earth, and
add new, till all which is Homogeneal be extracted
out of the Body. Then take this water exuberated
with the Fire of the Corporeal MERCURY, and
circulate it into a black Earth by continual
operation, and when the Water is so converted into
Earth, after it hath passed through all the other
Elements, take it and sublime it in a close Vessel
upon a gentle Fire, and what sublimeth to the top of
the Vessel, will be a Volatile Substance shining and
ruddy. This is that Thing which tempereth the
violence of our Mercury. Take this Substance very
pure, and add to it its Water, and distil the water
with it, and all which distilleth will be bright,
ponderous and unctuous, and is our glorious Mercury.

The following manner is very good, which is done
with white Sugar, which is brought in great plenty
from the EAST INDIES.

The Eighth Praxis of Sugar.

Take white Sugar-Candy which groweth in the East-
Indies in long Canals, and powdering it subtilly,
put it in a strong glass, and put to it the Acid
Spirit of Honey, very highly rectified; and by
Circulation, convert the Sugar with the Spirit of
Honey into a viscous Earth, which circulate, adding
more of the Spirit of Honey, till it be thin and
liquid in a palish colour. Take this and digest it
in BALNEO for thirty days, which time being passed,
open the Vessel, and distil off gently all the

superfluous phlegm, then change the receiver, and
close the joints exceeding well, distil a viscid
liquor of Sugar, which pour again upon what
remaineth, and distil the Liquor, cohobating so
often upon what remaineth, till it ascendeth with
the Liquor and leaveth the Feces behind, which keep
well, and take the Liquor, and circulate it by
itself for seven or eight days in a close Vessel,
then distil gently a clear and bright Liquor, which
is one of our MERCURIES; but what remaineth is thick
as thin Honey, to which put the FECES above
reserved, grinding both strongly in a Mortar, and
being very well mixed, put to this our distilled
MERCURY, and seal up the glass firmly, digesting
till our Mercury by digestion groweth red with the
Fire of its own Body, then separate it, and add new
to what remaineth, proceeding as above said, till it
will no longer grow red; keep what remaineth, and
take the MERCURIE, and digest them for two days;
then distil with a gentle Fire (luting exceeding
well) a white Liquor, which being distilled, put
your Vessel in Ashes, increase the Fire, and distil,
and there will distil a thick Liquor very yellow;
which keep, and rectifie very highly in Ashes till
it become clear, thin and bright; then take the
Feces and put them to those above reserved; mix
these well in a glass Mortar, and assate them, in
the beginning gently, then strongly, till after
various colours they become yellowish; which take,
and put to them drop by drop of the red rectified
MERCURY, and circulate; when they are dry, add more
of the red MERCURY drop by drop, circulating the
Mercury with this Earth so often, till it remaineth
moist by Circulation; then seal up the Vessel, and
digest it farther.

Now take Gold calcined, and pour upon it our Mercury, and distil the MERCURY so often upon the Gold, till the Gold remaineth white, with which SOLAR OIL the Medicine may be multiplied till it is most strong.

The Ninth Praxis.

Take the Minera of the red Earth, out of which separate a bloody and vaporous Humidity, which Circulate by itself for a long time, till from these is made one red fume, and another white; recifie each of these by itself, separating the superfluous and combustible SULPHUR. Then take the Body well known, and coagulate these fumes upon it by means of a dry Fire, and when this Mass is like Ashes, take it and distil, first a Liquor, which is LAC LUNAE extracted out of the Rays of the Central LUNA: and when it is distilled, take and rectifie it till it shineth like Luna in her bright Lustre (if it be first rectified upon SULPHUR VIVE sublimed, which must especially be here noted) this liquor is our vivifying Air, green and very powerful in Virtue to multiply LUNA, which is of kin to it. Take LUNA first subtilly purified by various labours, and precipitated into a Powder by a corrosive; of which, with this LUNAR WATER, make a thick and blue Oil, by a gentle cohobation of the said LUNAR MERCURY, upon LUNA so prepared. And here it is to be noted, that after every Distillation of the MERCURY upon Luna, the phlegm must be separated, which the Mercury during the Operation, by its own Virtue attracteth. Afterward, decoct that LUNAR OIL by a gentle heat of External Fire for 190 days, into a white shining Earth, which must be multiplied with the said LUNAR

OIL, till one part Tingeth five Thousand parts of
Venus into Luna.

This, Friendly Reader, is what I was willing to say
in general, by divers Praxes tending to one end,
which every one may easily understand, if he have
but the Knowledge of the SULPHUR AND MERCURY,
discovered by me in the First Book.

Chapter II.

I have sufficiently, in the foregoing Chapter, declared the PRAXIS in general, and openly enough to be understood. But now I will speak particularly of the Operations, and for the greater Illustration of the said very clear generals, that all who are Worthy and Faithful, may understand them. I know also, that for these PRAXES I shall have many ill-Willers and Reproachers, and chiefly those Philosophers, who persuade themselves, that the Tinctures may be made with a very little pains, in one Vessel, one Furnace, with one External Fire, and so deceive, with themselves, many others. But let these high nosed Scoffers Know, that the Philosophers Stone is a Thing of higher moment then they imagine. For it is a difficult Thing, and of deep Search to be understood; and of great labour to be accomplished; which they with me would acknowledge, if they apprehended the Operations of Nature.

But to what purpose are many Words? They had rather die than quit that Doctrine which is so radicated in their MINDS, by reason of their unlucky Interpretation of the Books of the Ancient Philosophers; but enough of this, now I will proceed to the PRAXIS.

The First Praxis of Mercury.

Although this Praxis at the first Sight, may seem ridiculous to many, yet it is true if it be understood. Take, IN THE NAME OF THE OMNIPOTENT GOD, of the best MERCURY, which must be pure, flowing, chrystalline, and very Serene, which you may very

well know, if you put it upon Silver; and after
Evaporation, it leaveth behind it a black spot,
which is a certain sign of Gold - or if you put it
upon a strong fire, it emitteth green and red fumes;
if it has these signs, it is good, and fit for our
Work; which you must purifie by subliming it divers
times, that at length by a long time, it may become
pure and neat, and freed from all corrosive and
phlegmatical Excrements, which take and pour upon
calcined Gold, mingling and grinding the MERCURY
with the Gold, till both are very well mixed; put
them upon a gentle Fire, evaporate the superfluous
MERCURY from the CALX OF GOLD, till you see it
appear of a red colour, then take it and grind it
subtilly in a mortar, and amalgamate it with new
Mercury, grinding without Fire, (which is to be
noted for this kind of amalgamation is made without
Fire) and when it is like a Paste, and eveporate it
again, reiterating this Amalgamation with new
Mercury, and Evaporation of the Said Mercury, and so
often till you see the Nature of this MERCURY to be
sufficiently introduced into this Gold, which may
easily be discerned.

This Gold is Mercurialized Gold, which take and
digest in a glass firmly closed for some time; which
being done, take it out and to reduce it proceed
thus, Take the Mercurialized Gold, and mix it with
the subtile and serene MERCURY in sufficient
quantity, and put the mixture into a good strong
glass Retort, which close well; and digest this
serene Mercury, circulating it upon the Gold, so
till you see no more to ascend and descend; and in
the bottom of the Vessel you will find the Gold
corrupted; of a black colour; which take and
amalgamate with NEW MERCURY, distilling, cohobating

and animating the Mercury upon the Gold, so often till the Gold is plainly reduced into a viscid Water, which is its Reduction, and requireth a long time; out of which reduced Gold now the Elements must be separated.

Here let all the Ignorant Sophisters and all Chymists be mute, who endeavour to make all Tinctures in a short time without any labour. These Idiots know not, that first there must be a Reduction of Gold into its first Matter; namely, a thick splendid and Viscid water abounding with the principles of compounded Gold. Secondly, the Separation of this reduced Gold, and Sublimation of the same most fixed Gold, before any profitable Tincture can be made. Verily an HERCULEAN Work, and most laborious And not known but by those who are Learned and Expert; which LULLY & GEBER, most subtile Doctors of ALCHEMY, do sufficiently shew; whose divers ways of working described by them, are to be accounted not Sophistication, as the Ignorant Chymic Mob persuade themselves; but for the very truth, let an Searchers of the Chymic Art, read the writings of the Ancient Sages, who have treated of ALCHEMY, from the most Ancient HERMES to the most Modern; they will find them all full of various Purifications, Reductions, Sublimations, Calcinations and the like, of the pure Substance of Nature; and also of the Distillations and Circulations of the Elements, which how laborious it is, no man who is Wise is ignorant of. For the pure Substance of the Mineral Nature, is the Stone of the Philosophers; which before its compleat Perfection, is a rude and undigested Mass, very much defiled by the Elemental impurity, which though it may attain the highest Purity (namely, that it become a meer

Fire, for the Stone itself is nothing else but meer Fire concentrated into one Thing) it requireth very great labour. But nevertheless (despising the Doctrine of the Ancients) these new Chymists and fraudulent gang, endeavour to obtain without any Labour and Industry, that which God hath given only to the Laborious; but they are deceived, and with themselves deceive many Persons who are Ignorant, credulous and Covetous of Gold; wherefore also for their sake, the true Art of ALCHEMY is esteemed as an unprofitable Juggle. For it is a far other work to divide and sublime Gold than they fancy; to the operation of which I will now apply myself.

Take the Gold reduced by Mercury and distil it and there will distil a Water of Gold; which being done, the Air of Gold will distil; and when this is over, then will distil the bloody and splendid Fire of Gold; and the Earth, as the fourth Element, remaineth in the bottom. Which take and assate it gently upon a gentle Fire, and when it is assated, take the distilled Air and pour it upon the Earth, and circulate till you have conjoined the Air and Earth. Then distil, and the Earth will distil in the Belly of the Air, and be suspended in it. This is a great Mystery, that the Earth should be suspended in the Air, and almost incredible, unless it could be ocularly demonstrated. Now take your Earth suspended in the Belly of the Air, and decoct it with Fire of Gold till it grow red, which is the great ELIXIR, which may be multiplied infinitely. O Wonderful Nature! Who permittest the Earth to be suspended in the Bosom of the Air, and also to inhabit with the Fire! Thou verily art wonderful, because thy Operations are wonderful. HERMES was very well acquainted with thee, when he said, "Separate the

Earth from the Fire, the thin from the Thick, sweetly, and with great ingenuity. It ascendeth from Earth to Heaven, and again it descendeth to the Earth, and receiveth the Virtue of the Superiors and Inferiors; and so the world is created." This Praxis of Mercury (which nevertheless is not that vulgar MERCURY which is sold by the APOTHECARIES which the ALCHYMISTS so wonderfully Torture, but another) is very good, and not only the Understanding, but also to the Working of which I could wish that all Worthy and Pious persons might attain; for whose sake I will subjoin the following Praxis for their greater illumination in the said Mercurial Praxis.

The Second Praxis of Mars.

Take CROCUS MARTIS, not vulgar, but calcined, and made purple red by Spirit of (also not Vulgar) VITRIOL, and dissolve that CROCUS in new Spirit will be very well Tinged; which decant, and circulate so lond, till the tincture by the operation of the Spirit upon it, beginneth to be VOLATILE, then distil, and again pour the distilled water upon the remainder, reiterating and cohobating so often, till the Tincture ascend together with the Water, which is the Tincture of MARS extracted from the Earth of our CHALYBS, which distil divers times by itself; then take what remaineth, out of which you have extracted the MARTIAL TINCTURE, and calcine it gently, out of which so calcined by gentle Coction, by a certain Art extract the Salt, which purifie very carefully, calcining, dissolving and distilling so often, till it is pure and passeth into red, as the innate colour of this MARTIAL MINERA, and

dissolve it in the said red Tincture, and put both
into a glass Vessel very well closed, and decoct
them both till they are fixed in a strong Fire. With
this Tincture of MARS (circulated upon Gold) you
must ferment, and again fix it; which if it be done
thrice (namely fermented with the fermental
Tincture) it will be a particular ELIXIR, whereof
one part fixeth 1500 parts of Venus into Sol.

The Third Praxis of the Fixed Body.

Take the FIXED BODY which you very well know (which
must be very pure) and dissolve it in water, and
when it is dissolved, put it presently to the Fire,
and distil gently the Phlegm, which cast away, and
when a certain acid water distilleth, change the
Receiver, and take it; and when the Vessels are
cool, pour this water again upon the remainder, and
dissolve it in a warm heat; and when it is dissolved
put it again presently to a distilling Fire, and
distil first the Phlegm, which cast away (for this
water in which the Body is dissolved, is like Spirit
of Wine, for it attracteth a Watery Phlegm) and when
the acid water distilleth, change the Receiver and
take it, as you did before. Continue this work of
cohobating and dephlegming for six times. Then take
the Matter which is in the Vessel, and is very much
corrupted, grind it subtilly, and put in in a glass
vessel, and sublime it by degrees of Fire, and take
this sublimed Mercury and dissolve it into Water
with its own Water, and when you have this Water,
take Luna finely laminated, and cast it into that
Water; and when it is dissolved, distil the Water
from the Luna, and what distilleth, pour again upon

the remaining Silver, reiterating so often till you see the LUNA turned into an Oil by cohobation of this Water upon it. Then take this Oil and put it to the Water with which the Luna was turned into an Oil, and mix them both well, and put them into a strong Vessel strongly closed, and put it into Ashes, and digest 120 days till they are fixed into a white powder, which take and dissolve in the Oil with its Water; and when it is dissolved, digest it after the same manner as you did before, reiterating this Work so often, till this Powder flow like Wax, without fume. And here it is to be noted, that after the Second Fermentation with the Oil, they are fixed in a shorter Time.

And this Tincture Tingeth much VENUS into LUNA.

The Fourth Praxis of the Green Lyon.

Take that Substance which in the first Book I named the GREEN LYON, for that is our Gold, living, and green, of a Saline Nature, produced by Nature out of the pure Substance of the Elements. Dissolve and congeale this, so often reiterating, till it floweth without fuming, which that it may be more easily brought to pass, this Gold in every Solution must be dissolved in its own Water. When this Gold floweth as Wax, take it and dissolve it in that Water in which our glorious Earth is resolved, and to every pound of Gold put half a pound of the glorious Earth dissolved in its own Water; and when the Gold is dissolved, digest it for 20 days in a warm place, and separate (the Feces which during the Circulation, fall to the bottom) by Decantation.

Then take this Liquor and put it into a Retort, and distil with a very gentle Fire as much as you can of the Phlegm; and when that is over, which you may discern by the taste, put out the Fire and cease Distillation, and when the Liquor is cold, weigh it, and if it weigh three pounds, take Vulgar Gold, and make a red Calx of it after the Vulgar manner, then free it from all corrosives, dry it, and put it into a circulatory (of this calx there must be half a Pound) and put to it the above said three pounds of Liquor. Then close the Vessel, and circulate in Balneo for 40 days, in which time you will see it will be all plainly resolved. Now distil, and what you distil, pour again upon what remaineth, reiterating till all what remaineth, be turned into Oil of a Golden Colour, thick and fat, which separate from its Feces by Decantation, and circulate it for 20 days, in which time it well become much Thicker and Fatter.

Of this one part tingeth 200 parts of LUNA into Gold. Because the glorious Earth is necessary to this process, and for the making of it, I have not given so particular a manner; I will also add the PRAXIS of this Divine Substance, for the sake of all those who seek this Art for the Honour of God, and the good of their Neighbour.

The Fifth Praxis of the glorious Earth.

Take the black Earth, which is also called Litharge, and put it into a Retort well luted, and distil it, and rectifie well all which distilleth, then take the CAPUT MORTUUM and powder it very well; for that is our black and obscure coloured LATON, which must be whitened, and its blackness taken from it, according to the Philosophers, who say, WASH LATON AND TEAR YOUR BOOKS, LEST YOUR HEARTS BE BROKEN.

Take therefore this Laton, and powder it in a glass Mortar, with a Pestle of the same matter, and when you have powdered it, assate it gently upon the Fire, then put it into a glass, and pour upon it that which you before distilled; then shut the glass and circulate the matter; and when it is circulated, distil it, and what you distil is our MERCURY. Pour this again upon what remaineth, and digest, and then distil, and when all is distilled, encrease the Fire, and out GLORIOUS EARTH will sublime, which is Our Fire Subtilized, our SULPHUR and our DIANA, which being so prepared, burneth Gold to Ashes, out of which is extracted the AURUM POTABILE, whose use is very great in restoring lost Strength.

The Sixth Praxis.

Take the GLORIOUS EARTH duly prepared, as much as you please, and calcine it gently, and put to this in weight of our Mercury, mix them well, then distil with a strong Fire, and urge as much as will ascend, and pour all which distilleth, upon the remaining part; and distil, and with a good strong Fire there will distil a thick, clear and gummy Liquor. Circulate this by itself 25 days; then rectifie it by itself four times, then dephlegm it; and this is our MERCURY Triumphant, in which dissolve GOLD, and by circulation make it an Oil, circulate this Oil till it be a dry matter. Imbibe this with the above said thick Mercury and fix it, which reiterate three times. Then take this Mass, and separate it from the Feces, as well as may be, by Sublimation, and what

sublimeth, decoct and imbibe without Triumphant
MERCURY, and fix it so often till it is fixed,
flowing and Tingeth. This way is short, and is very
powerful in transmutation of Metals.

The Seventh Praxis of Cinnabar.

Take red Cinnabar made of MERCURY VIVE by means of
Sulphur. Powder this very well, and pour upon it
Spirit of Salt very well rectified two pounds, and
mix them very well (and note, that in the mixing
them, the glass will grow very hot, which heat
cometh from the Internal Sulphur and Mercurial Fire
of the Cinnabar) then put it in Balneo and Circulate
for a long time. Then distil, and pour all the
Spirit of Salt which distilleth, again upon the
Cinnabar, and distil again, reiterating so often,
till the Spirit of Salt ascendeth very red as blood,
which distil gently in Balneo, and the Spirit of
Salt will distil, but the Tincture of CINNABAR will
remain, which keep, and pour the Spirit of Salt to
the remaining CINNABAR again, and extract the
Tincture as you did before, and if all the Mercurial
SULPHUR be so extracted out of the CINNABAR, and the
Spirit of Salt also distilled from the Tinctures;
then keep the Oil of CINNABAR, and pour again the
Spirit of Salt to the remaining CINNABAR, and digest
and distil at last with a strong Fire, and the
Sublimed MERCURY will distil like Crystal; which
dissolve in the Oil of CINNABAR, and mix both very
well, and distil them into one red Liquor which is
very precious.

The Eighth Praxis.

Make a Spirit of MERCURY VIVE, and when you have
well dephlegmed and rectified it, put it into a good
strong Glass, and put it to the Fire, and when the
Spirit is warm, cast in a little of the CORPOREAL
MERCURY, so often till it become thick, then
increase the Fire, and all will be dissolved, then
decoct this mixture till it be dry; then take new
Spirit of MERCURY and warm it, and cast in this dry
matter as you did before, and decoct, and so proceed
with new Spirit of MERCURY divers times; and if you
now see your Mercury like a Rose in the bottom, then
take this MERCURY and powder it subtilly, and put it
into a glass firmly closed, and digest it in 15 days
with a good strong Fire; then open the Vessel, and
put to it drop by drop in a glass Mortar, grinding
it, as much of the Spirit of Mercury as is
Sufficient, and the matter will be like a thin
paste, which digest seven days, then evaporate, and
upon the remaining MERCURY pour new Spirit to cover
it over divers Fingers breadths; then shut the glass
firmly, and put it to circulate 50 days in a good
strong Fire, and you will perceive the matter to be
fat; which circulate till it be again thin, then put
it to the Fire, and separate the pure from the
impure, and distil the pure, and there will distil a
very subtile Spirit, and what remaineth, will be
like Frogs Spawn, but whiter and more bright. Now
take the thin and distilled Spirit, and pour it upon
the remaining matter; digest, distil, and cohobate,
till this thick Oil ascend together with the Thin
Spirit, and when it is in the Receiver, it swimeth
above the Spirit, and is bright as Christal, which
Separate, and circulate by itself; then distil it,
and when it is distilled, circulate it upon LUNA,

85

and it will be a good work. This is what I have spoken particularly. But I hasten to what followeth, which is the practical way of our MERCURY upon the Calx of SOL or LUNA.

Temperate Water.

Take our GLORIOUS EARTH, which being calcined, circulate with our MERCURY, distilling till they become one Water, bright and clear, which is the TEMPERATE WATER.

The Use of the Temperate Water.

Take Gold and amalgam it with MERCURY, and evaporate the MERCURY, and calcine the remaining Gold, and edulcorate, till it becometh a Powder very spongeous, and purple red; which put into a circulatory, and pour upon it a sufficient quantity of TEMPERATE WATER, and work it by circulation, Distillation and cohobation till the Gold remaineth in the bottom in form of a viscid Calx, and separate the superfluous water. Now when this CALX is prepared, the Praxis is to be ordered in the following manner: Take of this GOLDEN CALX and our Earth ana[1], both being very well powdered; put to it drop after drop, grinding it, so much MERCURY as you see is sufficient, which you may easily see; then take this matter and put it into a glass Vessel and circulate 20 days, then take it out and put it in Sand, and distil our Mercury from the Calxes; and that which ascendeth after the MERCURY in form of a CRYSTALLINE POWDER, take and put to what remaineth, and dissolve it in Mercury, and distil, and it will

---

[1] In equal parts. -PNW

sublime which again put to what remaineth, and circulate, and distil, and there will remain the GLORIOUS EARTH of our GOLD and of our EARTH, conjoined by virtue of the Natural Fire, which take and reduce by our Mercury into a Tincture.

Thus far, Friendly Reader, by the Divine Grace I am come; and have, with a willing mind, instructed thee by divers PRAXES to make our Golden, Famous and so celebrated Stone, by which also, if thou beest ingenious, thou wilt see what advantage you may receive from them; and how difficult and Herculean a Work it is to come at. But that I may conclude this little Treatise, I will first advise thee to study to know these three principal things:

First, that you know the true Matter, which is only one Matter, out of which all our practical ways must be performed. This matter lurketh everywhere, and in all things: its name is SALT. This Salt you must know, before you begin any Praxis.

This Salt, tho', as I have said, it lurketh everywhere, yet it is not so commonly and openly found everywhere; for it is a hidden SALT, and lurketh hiddenly in all things, and for that reason it is called the CENTRAL SALT of all things:

Take this while it may yet properly be called an undetermined HYLEAL and hidden SALT.

Secondly, In the mixtion it is to be noted, that the water must predominate over the Body; for the necessity the Body must first be dissolved in its water, and turned into water, before the Body can corporifie the water, and which verily must be done

with very great caution, by little and little
dissolving the Earth; for the Earth is weak in the
beginning, and if you suffocate it with its water,
there will ensue a Sea of Confusion.

Thirdly, and lastly, the REGIMEN of the Fire must be
observed, which must be sometimes gentle, sometimes
strong, sometimes temperate, sometimes subtile and
vaporous, according to the Operation of the
Operative matter. The knowledge of this REGIMEN is a
great Secret, which above all things the Artist must
know, if he desireth the wished end; Which from the
bottom of my Heart I wish to all who are Worthy,
through our Lord and Saviour Jesus Christ.
AMEN.

FINIS.

THE THIRD BOOK
of
SANGUIS NATURAE

which was as yet not printed in English in this
present year

1705

Quid Reddam Domine

British Musuem MS Sloane no. **2037**

THE THIRD BOOK

OF

SANGUIS NATURAE

I have in my Two preceding Tracts so manifestly declared that Art which was kept so secret by the Ancient Philosophers, that they would not reveal it, even to their own Sons; both as to Theory and Practice: That any Man of a common Capacity, imploring first the Divine Assistance and Blessing, may easily apprehend it. In the first place it is most necessary to understand the Nature of the Matter, which the Artist worked upon, and its Purities and Impurities, and Active and Passive Virtues.

He who knoweth these will attain his desired end: For in the right Knowledge thereof consists the true Practice.

These I intend to reveal to all those who do not already know it: But yet with such Philosophical Caution, that the Readers hereof may not with unwashed hands touch these sacred Mysteries, which Almighty God from the Creation of the World to this day hath not revealed to any but his faithful Servants.

There are therefore Two chief Principles to be considered, namely Air and Fire, as an Agent, and Earth and Water as a Patient.

Of these four, two of which are only Visible, is made our Mercury, which is a perfect Creature, for it containeth the 4 Elements.

And here it is to be observed that Water and Earth are manifest to sight, and Fire and Air, are hidden, which is to be understood of the gross Matter. But when they are circulated by the Philosophical Wheel, the Fire and Air are made manifest, which is the Philosophers Mercury White and Red.

For in our Matter are Two Things, namely Purities and Impurities:

The Impurities are superficial, and must be separated being Faces which naturally adhere to the Matter. And if they are not well separated, the Fire cannot operate upon the Water, nor the Air upon the Earth; but pure Qualities act one upon the other. The Fire upon the Water, and the Air upon the Earth, and not separately, but altogether. For the Air cannot act upon the Earth without the Mediation of the Two others; namely the Fire and Water, and in like manner it is to be understood of the others. But our Water is not common Water, nor our Fire common Fire, for no Man is so senseless as to think that Water and Fire can be united. But our Water is of a Celestial Nature, and therefore our Fire loveth it.

The same is to be understood of the 2 others, namely Air and Earth; Vulgar EARTH is leprous, and aboundeth with excrements, but our Earth is chrystalline Green, and therefore it is called Green Gold. And our Air is unctuous and therefore containeth Fire. The Air impregnated with Fire, is

Celestial and Solar, and therefore hath the Virtue
of illuminating all Bodies, and of fixing and
exalting their Souls, of which all Philosophers have
spoken, saying, In Mercury is all that the Wise Men
seek. For in it are all the 4 Elements in the
highest Purity, and that is all the Philosophers
have desired. Hence it most clearly appeareth, that
the Matter of our Mercury is a perfect Body, and
containeth all things which are under the Sphere of
the Moon, that is to say, it containeth all things
potentially; for it containeth the Essences of all
things, and therefore is the first Matter of all
Things.

Out of this Matter purified (for it aboundeth with
Foeces) and duely prepared, and if I may say so,
made Spiritual, is prepared our Mercury, yet this
Matter in its first appearance is not so pure
(although it be potentially perfect) and yet
notwithstanding all the parts of our Stone are
hidden in it, and it is not apparent to sight, so
that it cannot show its Virtue, unless what is
OCCULT in it, be made manifest by Reduction, made by
an ingeneous Artist, so as it may become the Mercury
of the Philosophers.

If it be thus prepared it hath the Virtue to vivify,
illuminate and exalt dead gold, into the highest
degree of Tincture, which no other matter in the
World can do. And here apprehend us right, where I
call it a perfect Body, not because it is by nature
brought to a compleat Perfection, but because all
Perfection does both inchoate and end in it, and
gave it Power and Virtue to exalt Gold Radically,
according to the Will and Pleasure of the Omnipotent
who created it. For if it were not more perfect than

Gold, how could our Mercury extracted from it, exalt Gold?

Here it may be perceived that they all incur error, who endeavour to make Philosophical Gold out of Minerals, and imperfect Metals, as Vulgar Mercury, Sulphur or Antimony. These are not perfect bodies, because the Fire destroys them, and Gold itself is not by nature made so perfect, but that it may be destroyed either by Fire, Water, Air or Earth; therefore Gold needeth to exalt its Virtue.

This is that Miraculous Substance which containeth all Things, This is that Chrystalline Mercury always current, to which no Creature is altogether like: for in this, the Forms of Bodies do ascend and descend, as on a Ladder.

This is sometimes green, whose spirit if it be set at Liberty from its bonds, will reveal all the Essence, which before lay hid in the Centre. But these are secrets to be revealed to none, of which I dare not speak further.

I have somewhat deviated from my purpose, which was rather to explain the Agent and the Patient, the Male and the Female, the moist and the dry, which are Water and Earth in their Crudity, the two principle Pillars of our Glorious Mercury, must of necessity be perfected out of these two, namely out of the humid and dry nature. The Male and the Female, one must operate upon the other, so that the operation being finished, they might both become one again.

And Consequently that which was before of a lower form, is exalted and made our Mercury, clear and transparent, but how difficult this operation is, I will hereafter show.

The Earth laying in the bottom and purifyed, is subject to the continual operation of the water upon it, which when it once ascends on high, again it descends to the Bottom by drops, and by gentle Rarefaction is at last coagulated.

At first it appears dry, then moist and Viscous, then coloured like the Cameleon, which if it be acted upon, by the heat of a Bath, the Earth breaketh and sublimeth into a Water like Silver, which is our Glorious Mercury. This is the Star of the Wise Men, which being corrupted with the fiery Water, affordeth the Glorious Earth, which Wise Men call Diana.

But for the better understanding of this it is to be observed, that in this Earth called the Green Lyon, when it is purified, the form of the Water is manifested, namely the Soul, which is the Philosophers SOL, and when the time of operation is over, both of them come forth, the Fire of the Water, and the Air of the Earth.

For then their Occult becometh manifest, as above said, for the Water is appropriated to the Earth, hence the Air of the Earth stirreth up the Fire of the Water, which before by reason of cold was almost extinguished.

And therefore it is that Fire doth not appear in crude Water:

For the Water imprisoneth it, as the Blood doth enclose the Vital Spirit.

This Water is fat as a Subtile Oil, and therefore may be coagulated by the Earth, for this was Water before, and therefore is friendly to Water as abovesaid. And here you may observe the difference between Vulgar and Philosophical Waters: but the common Man doth not understand the Power of Nature, and therefore he always erreth.

This Water I have in my first Part called the Spirit of the Rock, and it is truly Rocky and Stony, and it is coagulated into the Stone of the Wise Men.

Hence the mistake of all those who work in Vulgar Waters and Spirits, is easily discerned: For these cannot be coagulated, tho' many do affirm that they may; but how learned these Doctors are, every days experience shews.

Hence I understand that Water which in the 2nd part I call Dew Water, Spirit of Wine, (whose phlegm distilled first) the Oil of Nature, the Spirit of Honey, the Crystalline and clear Mercury; this Water I have said is coagulated by the Earth, and by and with it is converted into an higher Form.

Here something is to be observed of which very many take no Notice, This Water when it is often distilled, and is made clear and Crystalline doth then leave behind it a White Earth, which before left a red one, and if it become thus far, then this Water is highly exalted in its property of easily coagulating, and the Colours in the Operation will sooner appear: because it is not so crude as it was before, for the continuance of the Colours proceedeth from the Crudity of the Matter and tho' it be distilled, yet it is in some sort Mature. For even distillation doth Maturate, altho' it elevated it, which the Chymists will not believe. For Maturation doth not much differ from Purification, and pure things approach more to maturity than impure things, which the Vulgar also Knoweth, tho' they are not Philosophers, and yet in this Art, it is a great secret, and the Learned Doctors do often seem to say the contrary, and from hence it happeneth that the sayings of philosophers are understood by few.

I have in some measure described the Matter and its parts, namely Water and Earth, of which the operation of our Mercury consisteth, now I will proceed to the Secret Operations which occur in. the Practice.

Water and Earth must be conjoined namely the Spirit, Soul and Body; these 3 which are involved in dark Feces must be purified, but this is rather a Regeneration.

For the Earth must be Chrystalline and flowing, the Soul Tinging and splendent, and the spirit serene and freed from all spots, but to arrive thus far is a difficult point.

The Intention of the Artist in this case, ought to be directed to this end that he may make the Water to dissolve the Earth: So that the Power of the one may be overcome by the other, namely that the Water may convert the Earth into its own Nature.

This is the best Operation for by this means the power of the Earth is augmented by the Virtue of the Water, that it may be the better coagulated.

But here it is to be observed, that this is an operation more difficult than the rest, which the Ancient Philosophers called the Conversion of the Elements, and said that in it consisted all the Art. Verily it is difficult to operate with the Elements, yet the Philosophers have proceeded so far, that even in their Glasses they can make a Coagulation of the Elements in this Praxis of the Body is the Agent, tho' it be dissolved into water, for in this Solution the power of the Earth is set free from its Bonds, and acteth upon the Water, for no Water can better be coagulated than that into which the power of the Earth is introduced, which is plain and evident.

Hence it is that in every Conversion of the Elements, this Practical Operation must always be observed, which proceedeth the real Transmutation of both, namely of the Earth and Water, as we have said, only that the power of the dissolving Water, may wholly contain the dissolved Earth.

So that before the real Transmutation there may be made a Solution per Minima and this is a great secret known to very few, particularly to those who know not the Philosophical Dissolution, and operation before the Transmutation.

Here are 3 things to be noted, first that the Earth must be porous, and light and not compact and solid; for so the Water doth better enter it with its subtil humidity and restoreth the Virtue, which was lost by Calcination, and augmenteth the innate heat, in a light, insensible and friendly way.

Secondly, that the Earth must be dry and the superfluous humidity, dulling the natural heat must be well extracted from the Earth by various solutions and Coagulations, for these humidities hinder the coagulating Virtue. And both these operations must proceed the Solution.

Thirdly above all, Care must be taken in the solution of the Earth when the same be dissolved gently per Minima, with a long trituration, so that the Virtue of the Water does not too suddenly lay hold on the Virtue of the Earth, for the Earth is weakened by divers Calcinations and purifications, even to Death; hence proceedeth a subtil Error of many in this difficult solution, in which even the least thing must be observed, that they do not debilitate the Virtue of the Earth by Addition of too much Water, and make a sea of confusion, a Pelagus Contserbationis.

This is what I have to say of Solution.

Now I will speak of TRANSMUTATION.

It is known that the dry elements mixed per Minima, with the humid, are easily altered and corrupted, for they are of one and the same nature, the Male and the Female, the one acteth upon the other, the Earth upon the Water, coagulating the Water into Earth, lifting the same up by a contrary operation, by means of External Fire.

Here the Seed of the Earth mixeth itself with the Seed of the Water, so as they can never be separated, for the seeds do here operate so long till they are overwhelmed with the darkness of the night and are inspissated into a Viscous Glew, whose Superficies is of a green and livid colour, as the signs of a sick and Languishing Complexion, for it is the beginning of Alteration.

Here the Spirit of the Volatile substance endeavours to free the Spirit of the Green Soul from its Corporeal Prison, and to join itself with it.

Here the Body so far as it is able resists the Spirit (least this Volatile and insatiable Enemy should devour its soul) and carries it into the Black Prison: for if this soul be freed from the Body, then Blackness appeareth as a sign of a good and right operation.

Here I speak something which a Vulgar Eye will not discern, therefore I will explain it, for it is a secret which few know. But why do I speak of secrets? All our Philosophy is secret, wherefore the Worldlings and Oppressors will never apprehend it.

There are many who cannot attain to any Blackness in the Work, and therefore err: for it is a very difficult work, and no man attaineth it, unless he knows first the natural Causes, therefore all the Philosophers in their Books do advise them to study, how to attain Blackness, if Blackness appeareth not in the Work, it is a sign that the soul which lyeth hid in the Body, is not set at Liberty nor dissolved; for that is the Medium which conjoined the Spirit, Soul and Body, and causeth that one hath Ingress into the other.

For it is the Workman of Nature, and as long as the soul lieth hid in the Inward parts of the Body, it cannot become a Medium of conjoining the Spirit with the Body.

Here give good heed, that you may amend this error, if you happen to commit it, But this Body is Salt, this. Soul lyeth therein enclosed, for it is that which must hold altogether, namely the Spirit, Body and Soul, and cause that one have ingress into another; but if this be not done, there is no Conjunction, which is like Black Pitch. In this Salt is also the Tincture which is not dissolved, but in this soul, hence it is called Putrefaction, but only a solution of the splendour.

For Salt cannot be putrified, therefore the putrefaction of the Philosophers is described after such wonderful manner, as that it is an Alteration into a better form.

Salt may be altered by the Fire of the living Water, which is the way to the Alteration of it.

Therefore also it is not to be wondered at, that when our Water is joined with our Earth, both of them are never at rest, but change from one Complexion to another, from one Colour to another, namely: White, Red and Black, and infinite other Colours, which alteration ceaseth not till they come to a Resurrection, and become much more powerful in Virtue then they were before, because when they are altered, they renew themselves to much greater strength, therefore Resurrection is not a new Creation but only an Exaltation. I have described this operation practically and largely enough in the preceding Books, which Lully hath delivered so obscurely, and that not without reason; for it is the Key of the Work as he confesseth in his Codicil and Testament, and that he had revealed to him by his Master.

But not only Lully hath hid it, but also Ripley, and infinite other philosophers being over envious, and so not willing to reveal so great a Treasure, lest Pearls should be cast before Swine. Verily I know none amongst them all that has so freely opened it with all its circumstances as I have.

For the Ancient and Modern Philosophers were as void of Charity as they were abounding in Riches.

You may abundantly see the Manuductions to this practice, in the first Praxis of the 2nd Book, page 73 of the Masculine Earth of Salt (rather Sol) and water of Dew. Also in the seventh Praxis, page 85 of the Vegetable Body of the Minera of SATURN, also in the Praxis of the glorious Mercury, page 87 of the Vegetable, Animal and Mineral body, and the ROCK WATER. Also in the eighth Praxis, page 88 of Sugar, and the Spirit of Honey.

But most clearly in the first of the Second Chapter in the 2nd Book page 94 of the Serene Chrystalline Mercury and Calcined Gold, in which every searcher may most clearly see the Solution of the Body in its Water, per Minima, and its Transmutation into another Nature, and also its full Exaltation together with its Water into the Mercury of the Philosophers.

Therefore this principal operation of our Mercury being known, it is also necessary that the Earth and Water, out of which such operation must proceed, be exactly known, and from whence they must be had, for these being unknown, the solution cannot be accomplished. I have said above that there is one Matter out of which our Mercury is extracted, which is divided into two parts, namely Water and Earth by distillation. See this Praxis in the 2nd Book page 82, in the fifth Praxis, where our Matter is mixed with Vitriol Calcined, as a Medium whereby the Water may better leave the Earth.

Also in the sixth Praxis, page 83 of Urine. Also in the ninth Praxis, page 91 of the Minera of the Red Earth. And in the 2nd Chapter of the said Book in the eighth Praxis page 109 of the Spirit of Mercury, withall requisit Manuductions, are perfectly taught.

The Earth and Water being now had, it is necessary first that both be purified, the Water is purified by seven or more Distillations, and is circulated afterwards into a Quintessence of the 1st Order. See this Operation in the 5th Praxis Book 2 page 82 of the Salt of Saturn, where the Water is rectified. Also in the Praxis of Urine page 83 where it is seven times distilled, also in the first Praxis of the Second Chapter in the 2nd Book of the Mercury, page 94 where this Water is 7 times sublimed in a long time, and freed from its Corrosive and Phlegamatic Excrements. This Water so prepared in our preceedent Book, is called the Spirit of Mercury vive - Spirit of Salt, Spirit of Vitriol,

Chrystalline and serene Mercury, Spirit of Honey,
the Water of the Rock containing the Soul of the
Elements, Aqua Vitae, of Red Wine, Dew Water, & etc.

But as to the Earth, that is purified by
Calcinations and Separations, by means of the
Natural Agent, for it must be calcined, digested,
made fluid, and reduced into a Porousness and
Redness, which are verily difficult Operations, the
Praxis of which see in the 2nd Book 2nd Praxis, page
76 of Calx viva, where this Earth is calcined to a
redness, by a dry Fire. Also in the 3rd Praxis, page
98 of the Salt of Tartar, where the Earth is
calcined into a Porousness. Also in the 2nd Chapter
of the 2nd Book, in the Praxis of the Mercury, page
94 where the Earth is amalgamated with the Water
and, is so often dryed till the Nature of the Water
be sufficiently introduced into it. Also in the
Praxis of the Green Lyon, page 103, where the Earth
is so often dissolved and purified, till it flows
like Wax.

This Earth in our Book is called Salt, Red Cinnabar,
the Green Lyon, Crocus Martis, Gold, White Sugar
Candy, Mercury, The Vegetable Body from the Minera
of SATURN, Salt of Tartar, Calx vive, & etc.

This operation of the Earth is difficult and Secret
and must be performed before the Water be joined
with it to make our Mercury. The Operation of which,
the Searcher has here openly discovered to him, and
may easily accomplish, unless he be an Artist of a
dull Capacity.

Although the pure Earth be joined with the pure
Water, per Minima, and transmuted into a higher

Form, yet our Mercury wanteth still another Operation to its Perfection. For after the Black and White Colours are past over, the matter must be sublimed and separated from some more subtil Excrements; namely from those which were not separated in the former operations.

This operation you may see, passing in the Practices of the first and second Book, and then our Mercury will be prepared.

But before I proceed further, I will reveal a great Secret. Our actuated Water before it be joined with the Earth, must be prepared into a Quintessence, namely, as it must be blue, green unctuous & ponderous, and not comminseible with common Water. This admirable secret lurketh in Vitriol and Urine: For the Golden Seed is in the Vitriol, and the Mercurial in Urine. Hence these two have great Affinity. Here take notice that the Philosophers do frequently say, that nothing but their Quintessence circulated for a long time, mixeth itself with Gold bodies, as I have said above. This liquor thus circulated abounds with a soul of an Earthly substance, therefore the Earth doth easily admit it. For after both are mixed, the Earth which before was hard black and Light and porous is now become fluid, white and heavy, and compact and of most easy sublimation.

And here note that as often as this Liquor is sublimed with the Earth, so often it is made more subtil and Vehement in Operation:

So that no liquor in the whole Mineral Kingdom is so subtil and Vehement: and which is wonderful it is

104

scarce discerned in the Glass wherein it is, by reason of its great subtilty and clearness. This Liquor altho' it be clear and Vehement in operation, and very volatile, yet it is very easily fixed by reason of the Fire which it carrieth in its belly, and therefore fixeth everything with which it is mixed.

I should say more of this glorious substance, if it were lawful to cast Pearls before Swine. All mankind loveth nothing more than Gold, and since that cometh now in my way, I will teach the way of making it, out of the baser Metal: which making many have called Transmutation, and that rightly, for it is a Transmutation of a baser form into a better, without hurting those things which make the Tincture, for where light is introduced, Darkness ceaseth, therefore these 3 thing are to be considered, viz, the Confection, the Transmutation and Introduction.

The Ancient Philosophers out of a zealous Piety and Charity made light accomp of this Operation, searching a Medicine healthful to Mankind, which is quite contrary to men of our age, for Gold is their Idol, and they dispise Spiritual Treasures, as the rest of the World do.

The Mercury of the Philosophers being now very well prepared, rectified and glorified, (for it also needeth Glorification) and the Glorious Earth being in the same manner prepared they must be mixed accordingly to the proportion of Nature with the soul of Gold by divers operations, in a long time, till all thicken as may be seen in the first and second book: then take this Matter (for in it lyeth all which is necessary to this Work) and Work it till it grows fulgent as a Pearl. Then decoct it to blackness with a gentle and moderate Fire, then into Whiteness, first passing through other colours (which I will not here mention being sufficiently described in other Books) and lastly into Redness.

Work this Stone imbibing it by little and little with the Soul of the Gold, till it be absolutely perfect, red and fulgent.

This operation is plainly described in the first and second Books, and therefore I will not here describe it more plainly, since I have done it already; not for the sake of such as love Gold, and of those whose Heaven is in this present life, but for the sake of good men and well deserving in this Art, and of those whose Treasure lyes not in the Things of this World, but in things spiritual and Celestial expecting the last coming of one Blessed Lord and Saviour Jesus Christ.

I have been more prolix in this Discourse than I intended and therefore I will hasten to a Conclusion; but I fear I make music to the Deaf, for in this Age the World is Governed by opinion. So many men, so many opinions, which many times are as absurd, as impossible: So that some affirm White to be Black, and yet these are the great Doctors, and wise Philosophers, and oftentimes the Expounders and Commentators of the Adepts.

Vain Ostentation! And foolish Ignorance of Man! O miserable and empty Philosophers, fit to be banished and excluded from the Conversation of Men, and these are to be shunned more than poison for they are the

Corrupters of true Natural Philosophy, but enough of these Men.

Now I will speak of Urine and Vitriol, in which that so much wonderful secret of our Quintessence lyeth hid; which few know, and very many will not believe, for it is hidden by Divine Providence least the Ignorant and Vulgar should know it as well as the Wise-men.

Vitriol is a salt, which (not without weighty Consideration) is taken into the Composition of the Quintessence: For it is.a Salt of the Nature of Fire, full of Tincture, red and White, and often black, perservering in the fire, of a Vegetable Nature, and therefore it is green and yields a green Tincture in Vinegar. Such is our Vitriol, much different from common Vitriol, whose qualities are not so noble and powerful, nor so fix and Tinging as the qualities of this Golden Vitriol.

For it hath so wonderful a Tincture as is scarce credible, because of the great Projection, which it makes-upon Venus which it turns into pure Gold.

This is such a Tinging Substance as none in the world is equal to it; this Vitriol is found everywhere, and is of most vile price:
It is sold everywhere, and no Creature can live without it; for in it is shut up that ethereal Nectar, the Nutritive Substance of all things, and here also fixed, that it therefore may operate so much the stronger. Behold now I have clearly described it, and if you do not apprehend me, you ought to confess yourself ignorant, because the Description of this substance is so clear.

This Vitriol as I have said, hath a golden seed, therefore it is green to the sight and in Virtue, and therefore called the Vegetable Saturnia: This Greenness is visible till it is set at liberty from its Bonds: For when it is set free it is red and no longer green and consequently more perfect.

This red spirit is the principal, part of our Quintessence, for all that is fixed Volatile and dissolved by it (the 5th Essence) is done by this circulated Spirit. Reader, if you know it keep it secret, for it is the first step to our secret Fountain, and if this green and red Spirit of Vitriol, be unknown, this Fountain is not found.

As to what concerned Urine, I will make no large description of it, for all men know what Urine is, yea boys and girls know how to discourse of it. This substance which now I call Urine, because it is so vile, and to be found everywhere, and is in the power of all, therefore it is lightly esteemed. But if Men knew the Virtue and power of this Urine, they would seek it to the end of the World.

Therefore Wisemen knowing its inward fiery and hot, and its outward cold and moist qualities, have called it by the Name of Urine of Saturn, and this not without weighty reason, which it is not lawful to reveal to the unworthy. Therefore Geber, speaking of Sulphur hath partly hinted these Noble qualities which Urine containeth, saying:

By the Most High God it is that which illuminates every Body, and it is a Light from a Light and Tincture.

Verily these are Weighty Words, which do shew the splendour; Light and Tincture of this Urine. For it is known to Vulgar Chymists, that out of the Urine, common to Men and Beasts (which nevertheless differs much from the Urine of Saturn) that lucid liquor called Phospherus is prepared; and why should not out of our Urine, (which inwardly is almost wholly fiery) be prepared this Lucid and splendent Salt, by the Philosophers commonly called Diana.

This lucid Salt in our Quintessence is of so great Efficiency, that it vivifyeth and illuminateth dead Gold, this Salt before its preparation is meerly phlegmatic, stinking and black, (for Urine putrified groweth black) but after its preparation and fiery circulation it is sweet smelling, most white, flowing and splendent; sometimes also it is viscous as Oil (but seldom) for then it is brought to the highest degree of perfection, and hitherto few have attained to it. This Oil when it is distilled, gives a Light, and swimmeth upon the Quintessence, and is endowed with so much purity and subtilty, that it can scarce be separated from our Glorious Liquor. But it is separated by a gentle abstraction of the Quintessence, leaving this subtil Oil in the Bottom of the Cucurbit, which must be once more rectified, that it may become more pure. This Oil hath wonderful Effect in Transmutation upon Mercury, for it easily flows, and is of most subtil Penetration. These things lye hid in Urine, a subject so vile and despicable. Reader, here you see, that in Vitriol and Urine, so wonderful a Secret of our Quintessence lyeth hid: For this liquor participates of both Natures, the Sulphureous and Mercurial part.

These are those two Substances, which in my
foregoing Books I called Sulphur and Mercury, and
without which nothing is performed in the Work.

Study therefore to know these for they are the Roots
of the Physical Works, and now I will conclude this
discourse.

Study very diligently what I have said scatteringly
in these three little Tracts, namely to know the
force of Nature, it active and passive power, and
lastly to know thyself. These are the principal
materials of this Art, which if you take for your
Foundation, you will work well; for knowing the
force of Nature, you will know the possibility:
Knowing the Operation, you will know the theory;
knowing the Active and passive Power, you will know
the Practice: And knowing yourself, you will find
all which you seek for.

These are the Mirrors in which you will see all
Things Sublunary, and believe me upon my word,
unless these be your Foundations you will find
nothing. For they are the way which leads to the
Castle of Wisdom: Behold they are difficult ways for
Worldlings, into which it will be impossible for
them to enter. Hence cometh the small number of
Adepts, and multitude of false Philosophers, for the
former overcome the difficulty with humility, and
become Victors, the latter with pride contend for
lighter things and are overcome.

Therefore pray unto God thy Creator, who is Almighty
and Merciful, and will not forsake thee, if thou art
of a good heart and mind, and will assuredly prosper
thy undertaking.   FINIS.

N.N.N. OF LULLIUS

The preparation hereof consists in uniting a
Volatile Salt, and the Oleosum Mellis, as in Sanguis
Naturae: And all such Adepts call it by the Name of
Rebisse.

Others join to the SPIRIT OF URINE either rectified
SPIRIT OF WINE, or SPIRIT OF HONEY. Lullius calls
this Mixture of Spirits, Sphera Ciebia, without
which mell (?) is of no use to us.

The Preparation of this Sphera Calica he describes
according to the letter in Libro Experimento 4:
(viz, about the beginning of that Book). The use of
this N. N. N. is most amply described in Testamento
Novisgimo, wherein, according to the Letter, are
contained the greatest secrets of Lullius, so that
thereby he tingeth imperfect Metals give juvamento
ignis.

Julius Book de Quinta Essentia, he digests his N. N.
N. in his Sphaera Calica, about a month, in which
time the Oleoium Mortificatum will be revived and
swim at the Top in the form of a Celestial or Blue
Colour, which he calls his SPIRITUS VINE
PHILOSOPHICI.

Guido Monteuor (a Grecian) digests this mixture for
6 weeks to the same purpose. Paracelsus for 2
months; but this difference is occasioned only by
the different Regiment of fire, being more or less
intent and the different Processes and the Vessels
they used.

Christopherus Parisiensis in his Appendix calls this mixture of Sphaera Colica and Mel, Chaos Philosophical, or Prime Materia Metallox, and according to the letter describes its use both as to the restoring and meliorating of metallic and human Bodies.

Julius Elucidarium he distills the N. N. N. as Lullius does in the beginning of his Testamentum novissiume. The Author of Sanguis Naturae, Ripley and others, distil it into a Red Oil, which afterwards is rectified into a clear liquor called our MERCURY, which mixed with his Sal Amoniaecum Vegetable, produces a Menstruum Vegetable Simplex, which the Author of Sanguis Naturae calls his Triumphing Mercury.

It is to be observed that Chr. Parinensis in his Appendix, as also in his Elucidarium sub Capile de Rubeo, bids us wash the N. N. N. twice in its own phlegm, that the Aridium may be the better separated from the Pix nigra.

Also it is to be observed, that also Lullius and Paragenus work immediately upon this Pix, that the Author of Sanguis Naturae bids us to digest the said Pix, till it be turned into a black Earth.

In a word he who understands the Pix Nigra, or N. N. N. of Lullius, unto him are opened the practical parts of all the Adepts to be understood according to the Letter.

July 8th. 1705. The fluxing Powder for Tincture.

Rx. 8 parts of Nitre, 1 part of Quicklime, viz, such as is made of marble or Stone, as also Eluis. Powder them both and mix them in a Platter, pour water upon them, and by stirring the mixture and take the Quicklime; then set this mixture in the Sun or in a warm place to dry, and powder it. The use followeth. Take three parts of this powder, and I part of such Tin as naturally contains LUNA: but if such cannot be had, take JUPITER and put LUNA to it; granulate it in water, make S. S. S.[2] with this JUPITER & Powder, in a Crucible, put the same into a Founders Furnace, till it flows like WATER, let it cool, break the Crucible, and you will find the Silver at the bottom, in form of Regulus Double refined, and all the JUPITER reduced into a scoria at the top. Note the lightest and whitest Quicklime is the best.

Observations.

1. If the fire be not strong enough, and the Compound not well melted, the LUNA cannot subside, but will lodge in small particles or grains within the body of the Compound, which by grinding in a Morter the Scoria and the JUPITER may be washed away.

2. This fluxing Powder, has this extraordinary quality, that it separates the Silver most highly refined from all Metals and Mixtures, as for example, should the JUPITER you use contain besides the LUNA any imperfect Metal as VENUS or Mundie (an Ore very near to JUPITER) or you should add some VENUS on purpose, to the JUPITER, this powder would

---

[2] Layer upon layer. -PNW

free the LUNA from these impure Metals, and turn
them all into Scoria.

3. The JUPITER contained in the Scoria, is very
proper for Pavements, walls in Gardens, etc. And it
is probable, that it may be made ignible as MARS,
VENUS and Brass, and so useful for Fire Grates,
great and small Guns, all looking like silver LUNA,
and not to be distinguished from it except by the
Test.

FINIS.

THE QUINTESSENCE OF THE BLOOD OF

NATURE

OR THE BOOK CALLED SANGUIS NATURAE, PURGED FROM ALL
SUPERFLUITIES OF WORDS, THAT IT IS BECOME
INTELLIGIBLE BY EVERY DILIGENT ENQUIRER

1.   The whole secret of our Art consists in the manifestation of the Light of Nature, which is imprisoned in all Bodies.

2.   This manifestation of the hidden Light cannot be performed but by the light which is first manifested in our Philosophical Matter.

3.   To understand and to do this, we must know that the light of Nature, which is the form, the life and virtue of everything is one and the same in all created Beings.

4.   But may fitly be divided into a Volatile and fixt, or universal and particular Light.

5. The Volatile or Universal light flows from that great Ocean of Light, the Sun, into the Stars, Fire, Earth, and Water.

6.   The fixt or particular Light is more or less hidden in every Elementated Being, into which the universal Light is magnetically attracted by the particular light in them imprisoned for its nourishment, multiplication, Conservation & etc.

7.   Between these Extremes the Light is another as a medium, which is neither Universal nor Particular,

Volatile nor fixt, but participates of both, and is thus generated, viz, when the Volatile Light of Nature descends from its Father the Sun, and assumes a Body in the Air, in uniting itself with the supercelestial Waters, then it becomes clouded, and remains an undeterminated substance, not being as yet attracted by any of the magnets of our three Kingdoms, and consequently not specified. This in our Art is called the first matter, Sanguis Naturae, Ignis Naturae, Balsaums Naturae, Semen Universale, Magnesia, Draco Viridis, & etc.

8.   This our Philosophical Matter is but one only substance, unto which nothing can be compared in the whole Universe.

9.   And although this matter is neither Animal, Vegetable nor Mineral, nevertheless it contains the Virtues of them all.

10. Especially it contains the properties of SOL & LUNA, whence it is often called our SOL & LUNA, or Gold and Silver.

11. It appears in the form of a Salt.

12. In the knowledge of this wonderful subject consists the whole Practice.

13. In our Practice this only matter is considered, as divided into 2 parts, of which one is called the Water, the moist volatile, mercurial part or Agent; the other, the EARTH, the dry, fixt, sulphureous part, or Patient.

14. These 2 parts of our matter are mentioned in the Title page, and called the Sanguis & Solar congealed liquors, the masculine Earth of SOL, and Water of Dew, the Vegetable body, and Rock Water, Sugar or the SPIRIT OF M; calcined Gold, Serene MERCURY .& etc.

15. It is said that this admirable Secret lyes hid in Vitriol & URINE. Without these 2 nothing can be performed in this Work.

16. It is worth our trouble to study the nature of them both. We will begin with the Water, and the 1st. thing which we must learn, concerning the same is that the Philosophical Water hath a great love or sympathy to the Philosophical Earth, since it is prepared out of the Earth, and is afterwards to be joined to it.

Out of this matter purified (for it aboundeth with faeces) and duely prepared, and if I may say so - made spiritual - is prepared our MERCURY.

I have somewhat deviated from my purpose, which was rather to explain the Agent and Patient, the Male and Female, the moist and dry, which are Water and Earth in their crudity, the two principle Pillars of our Glorious MERCURY, one must operate upon the other:

So that the operations being finished they might both become one again.

It is known that the dry Element mixed per minima with the Humid, is easily altered and corrupted, for

they are of one and the same Nature, the Male and
the Female & etc.

It is necessary that the Earth and Water, out of
which this operation must proceed, be exactly known,
and from whence they must be had, for these being
unknown, the solution cannot be accomplished.

I have said before; that there is one matter out of
which our MERCURY is extracted, which is divided
into 2 parts, namely Water and Earth by
distillation.

The Reduction must be made by a certain contrary
liquor (as to FIRE or SOL & LUNA, hidden in our
matter) for SOL & LUNA which are secretly in our
matter, and rule powerfully in it, are not reduced
so as to appear to light, unless this Reduction be
made by a contrary, which is a Menstruum or most
subtil Vapour (or Water) penetrating and resolving
containing in it AIR, FIRE & WATER separating the
pure from the impure, and is 1st. extracted our of
our Minera, or Philosophical Matter.

That SULPHUR which we call the Green Lyon is the
FIRE of Nature, which lyeth hid in the centre of our
subject, understand Salt, and is there detained,
shut up in a strong Earthly Prison, unable to exert
its force, unless by its Associate it be set at
liberty from its fetters, as that it may come out
together with its companions: It is not easily
dissolved except in its own liquor, for it is its
Companion, its Aery Companion.

This Green Gold is clothed with a foul Garment,
which must be separated by dissolving, by help of

the MERCURY of Gold 1st. extracted out of the said Green Gold. It dissolves nothing but the Golden nature of Gold which is of its own nature. This Water of a wonderful sympathy loves the Rock from whence it issued.

Our Solar EARTH needs the Water which is its female. Take this Solar, golden and ruddy Earth, and add to it the Water of Dew, which is its Wife and Mother, for this Earth is generated by Dew, and the Water will be impregnated with the golden seed of the Male.

17. This Water of Philosophers ought to be considered in its several qualifications, natural as well as artificial: we will begin with it, as we receive from nature, and is not in the least prepared by Art, and then this Water is called in plain Terms URINE.

18. That our Adept by Urine or our natural WATER, understands nothing metaphorically but common URINE, appears from hence. 1st. That he speak of URINE which the very boys and girls know, and it is in the power of all mankind. 2nd. Out of which the Philosophers MERCURY is made.

3. That we ought to look for this URINE at the end of the World: Viz, the Microcosme.

4. That he bids us to learn to know ourselves in order to find out the Materials of our Work and when found out, to give God thanks for the Wisdom and power of God has granted us by it.

5. That the operations done with this URINE agree with those of the common URINE as will be seen in the Practice. Yet it is to be observed,[3] that the Adept speaks of a peculiar and not common URINE for he says, if men knew the Virtue and power of this URINE, and etc. All shew the splendor, light and Virtue of this URINE, why should not our of our URINE (spoken to opposition to common URINE) wise men call it by the name of URINE of SATURN. To this is answered that in the above mentioned places in Sanguis Naturae, there is spoken of a two fold URINE, simple and compound, or Natural and Artifical; the simple and natural is the common URINE here treated of, the compound or Artifical is the common URINE mixed with the Philosophical SATURN and is called URINE of SATURN.

20. This URINE or natural water of the Adepts is further to be considered, in its several artificial qualifications, as 1st. it must be putrified, and then inspissated to a black Salt. Take URINE putrified and inspissate it, out of which so inspissated make a Black Salt, which is an Animal Salt, & etc.

This Salt before its preparation (inspissation) is merely phlegmatic, stinking and black, for URINE being putrifyed, grows black, but after its preparation and fiery asculation, it is sweet smelling most white and splendent, Before it is set

---

[3] This is section no. 19 in this current series and is run together in a single paragraph with no. 18, therefore I have typed this as it stands. This whole commentary on Sanguis Naturae appears to not be written by the Author. -D.H.

at liberty, it is rude, vile, abject undigested Mass, which is also found scattered in the Earth, (rather contained in the Earth, viz, in the inspissated URINE or black Salt) out of 100 lb. whereof scarce 1, or 2 lb. of which is pure, the Soul, Fire, Oil & etc.

There is only one salt useful to us, a pontic fiery, bitter and Mineral Salt of a Saturnine nature out of which this famous liquor is extracted, which is of so great moment; it must be distilled and rectified, for in this there are caustic viscous and bitter salts, all which must be separated, otherwise they prejudice the Work.

This thin and viscous substance Urine, which we also call our MERCURY, doth abound (in its natural condition) with many aereal and viscous Excrements, which savour of the nature of fountain Water; but there are others (in its Artificial quality) which are of a greasy, oily and fat nature, and are the corresponding and caustic Fires of a Sulphureous nature, which also must be separated by distillation.

21.[4] The properties of common distilled URINE are expressed in the following Places to be corrosive, pontick, bitter, sharp, white, serene, ponderous, ethereal or very subtil, that which we desire to perform, ought to be done with our corrosive, pontick, fiery, precious, fetid, bitter and sharp

----

[4] Note: there is no indication of where no. 22 begins or ends, but goes from 21 to 23; and so there you have it, just as the original is, (for what it is worth). It appears that the numbering system was inserted after the writing was completed, and sort of at random, and therefore of little consequence. -D. H.

MERCURY, and is called by the Names of all sharp and corrosive liquors.

Our Golden MERCURY is a white, serene, ponderous, acid & pontick liquor, of an etherial substance which is that so celebrated Animal, Vegetable, and Mineral Mercury which & etc.

Nothing does more destroy these (heterogeneal and corrosive) homogenial parts, than our pontical and corrosive MERCURY, by reason of its fiery Nature. NB. per se; After we have drawn out all the stinking and menstrous spirits from the Mineral Body.

Thus our Calcination is the Augmentation of the innate FIRE, and the highest purification of the body which is done by our pontic Water full of FIRE which burneth and mortifyeth the Body. Our humid MERCURY which containeth the fire of the Elements is extracted out of our only Minera by force of External fire.

This is made with the pontick WATER full of living FIRE which alone is capable & etc.

Now let us return to our WATER, which is a certain WATER very subtil, and precious, acid, fetid, corrosive, and sharp, which the Ancients hid under the Name of Vinegar, as also of other acid and fiery liquors, as of AQUA FORTIS, Vitriol, Alum, Salt peter, and Sal armoniac, which WATER is called Acetum acerrimum, because it is very sharp and acid.

This WATER is called Aqua vitae and vegetable and Animal spirit of Wine, strong Vinegar, saturnal WATER, and many other names; as Rock water, Argent

Vive, a fume, the tinging celestial SPIRIT, incombustible FIRE, Wine Vinegar, Saccus Acatia, SPIRIT of Wine, Temperate WATER, Luciferous Virgin, all the names signify this WATER.

This distilled URINE must be very well purified or dephlegmed by 7 or more rectifications that it may become pure-serene and crystalline, and if you let somewhat thereof drop upon red hot LUNA, it leaves-a black spot behind, but cast into the FIRE emitteth green and red fumes; in the distilling it leaveth a White Earth, and is freed from its corrosive and phlegmatic Excrements. Take URINE, putrify and inspissate it, out of which so inspissated make a salt, which is an Animal Salt, distil this in a strong Retort, and what is distilled rectify 7 times.

This water is purified by 7 or more distillations & etc. See this operation in the 1st book of the Salt of SATURN, where the water is rectified. Also in the Praxis of URINE where it is 7 times distilled. Also in the 1st Praxis of the 2nd Chapter in the 2nd book of MERCURY, where this water is 7 times sublimed, in a long time and freed from its corrosive and phlegmatic Excrement.

This water when it is often distilled, and is made clear and Chrystalline, doth then leave behind it a white earth, which before left a red one, and when it is come thus far, then this water is highly exalted in its property of easy congealing. Take the best MERCURY, which must be pure, chrystalline and very serene (made so by 7 cohobations) which you may very well, if you put it upon silver made red hot, and after evaporation it leaves behind it a black spot.

The SPIRIT of URINE thus rectified is called by the following Names. This water so prepared in our proceeding books is called SPIRIT of MERCURY VIVE, SPIRIT of Salt, SPIRIT of Vitriol, Chrystalline and secret MERCURY, SPIRIT of Honey, the Water of the Rock, containing the soul of the elements, AQUA VITAE of red wine, dew water and etc.

I understand by that Water which in the 2nd. part I call Dew Water (whose phlegm must first be distilled) the oil of nature, the SPIRIT of Honey, the Chrystalline and clear MERCURY.

This water I have in my 1st part, called the SPIRIT of the Rock, which is truly Rocky and Stony, and is coagulated into the Stone of the Wisemen. I called it Water of Dew, Rock Water, SPIRIT of Honey, Serene and Chrystalline MERCURY, the best MERCURY SPIRIT of MERCURY VIVE well dephlegmed and rectified, SPIRIT of Wine, AQUAE VITAE distilled from Wine, distilled Vinegar, Water of Rock.

This stony SPIRIT is white, acid, and containeth the Soul of the Elements & etc., the Acid SPIRIT of Honey, SPIRIT of Vitriol, Water of Dew which is its wife and Mother.

23 & 24. It is to be noted that not only those names above mentioned, signify and do belong to the SPIRIT of URINE, but these also, and all names whatsoever given to water and liquor, which is to be used or joined with the Earth, the other part of our only

Philosophical matter, and that all these Names signify one and the same SPIRIT.

25. But the most proper name given to the SPIRIT of URINE is Water or SPIRIT of the Rock; for the SPIRIT of URINE is truly Rocky or Stony, and often coagulated into stones or Chrystals, white, and containing the soul of the Elements.

26. It was said if our only Matter was to be Divided in to 2 parts, WATER & AIR, and in that these two parts were called Vitriol and URINE, we will proceed to the other part, the Earth or Vitriol.

27. This VITRIOL is not common but Philosophical Vitriol. This admirable secret lurketh in VITRIOL and URINE, for the golden (sulphureous) seed, is in Vitriol, the Mercurial in URINE, hence these 2 have great affinity.

Now I will speak of VITRIOL and URINE in which that Wonderful secret of our Quintessence lyeth hid, which few know and very many will not believe; for it is hidden by Divine providence, least the Ignorant and unworthy should know it as well as the Wise Men.

Vitriol is a salt which not without weighty considerations, is taken into the Composition of the Quintessence, for it is a Salt of the nature of fire, full of Tincture, red and white; and it is often black, persevering in the fire, of a Vegetable nature, and it is green and yields a green Tincture in Vinegar.

Such is our VITRIOL, much differing from common
VITRIOL, whose qualities are most noble, and
powerful, nor so fixt and Tinging as the qualities
of this golden VITRIOL, for it hath so wonderful a
Tincture, as is scarce credible, because of the
great projection which it makes upon VENUS, which it
turns into gold. This is such a tinging substance as
none in the world is, to it.

This VITRIOL is found everywhere and no creature can
live without it, for in it is shut up that Ethereal
Water, the Nutritious substance of all things, and-
here also fixed, that it may operate so much the
stronger.

Behold now I have already clearly described it, and
if you do not apprehend me, you ought to confess
yourselves ignorant, because the description of this
substance is so clear.

28. This VITRIOL, as I have said, hath a golden
seed, it is green to sight and in virtue and is
called the Vegetable Saturnia. This greenness is
visible till it be set at liberty from its bonds,
for when it is set free it is red and no longer
green, and consequently more perfect & etc.

Here you see, that in VITRIOL and URINE so wonderful
a secret lyeth hid, for this liquor participates of
both natures, the sulphureous and mercurial part.
These are the two substances which in the 2 former
books I called SULPHUR & MERCURY, and without which
nothing is performed in the Work.

29. Having learnt that the Philosophers have but one
only matter, and divided into 2 parts, and also how

one of them ought to be purified, it remains now to know how the other the dry part is to be cleansed from its impurities.

This purification of our Earth will be plainly taught in the re-uniting of the said divided parts of our only Matter in which conjunction of these 2 principles consists the whole preparation of the philosophical MERCURY.

Water and Earth in their crudity are the 2 principal pillars of our glorious MERCURY; for the MERCURY must necessarily be perfected out of these 2; viz, out of the humid and dry nature, the male and female, one must operate upon the other so that the operation (purification) being finished, they might both become one again, and so that which was before of a lower form is exalted, and made our MERCURY clear and transparent.

I have in some measure described the matter and its parts, viz. Water and Earth, of which the operation of our MERCURY consists.

Now I will proceed to the secret operations which occur in the Praxis, Water and Earth must be conjoined & etc.

30. The operation of our philosophical MERCURY is also divided into 2 parts, in the preparation of our MERCURY, and of our Glorious and Triumphing MERCURY, the 1st. is simple and an essence, the other compound and a Magisterium.

31. The preparations of our simple MERCURY consists in these 4 operations, viz, to make a black and red Earth, and a red tincture of a red Spirit.

32. The preparation of the Black Earth is described in general in the Praxis of the glorious MERCURY: where you will find these words. Take our corporal MERCURY & etc. circulate it into a black earth by continual operation.

33. For the better understanding of this process it will be necessary, that we examine every member thereof more particularly.

1. There cannot arise a doubt. or scruple which our Adept means by our corporeal MERCURY, Animal, Vegetable & Mineral; since once for all he assured us, that we ought to understand always that only philosophical matter, let him call the same by many and different names. We take the matter which in the title page is called the solar congealed liquor of Nature, and as above is called by many other names.

2. This our Matter is said to abound with many impurities, from which it ought to be cleansed. It is not the whole substance of the 1st. matter which endures the fire, but only its pure parts, wherefore it is necessary in the 1st place, to purify the matter and take away the sphere of SATURN, which cloudeth the SOL & LUNA (illo JUPITER splendour) before they can despise the fire. This operation is called by philosophers a destruction of the compound, for rebirth.

-Finis-

John de Monde Snyder's
Universal and Particular Processes.

Translated from the Ehrenthal Manuscript.

The principal obstacle and most prominent corner stone whereon most persons have stumbled in the Universal Work is: Primo la materia, secundo the menstrum thereof and tertio the true destruction of the metals, namely, into their three principles viz.

Now how these things all follow each other, and must be taken for the high work of the Universal Tincture, will con ajuto dell Omnipotente be clearly described and noted in due order in th following, as is done in the actual and accomplished work & wrought by hand.

Firstly according to the description of the true Philosophers, among whom John de Monte Snyder, should be most readily and on account of his clear exposition most eminently considered, our true and genuine Master is the old dragon, and white Eagle, Man and Woman, two Mercurial Waters, Dua argenta viva, efconpin diversi nomi, but indeed it is with its own name Antimony and the highly clarified Sublimate. Of these two can the high work be made, to shorten the time however these two are animated & fermented with metallic ♁ & ⊖.

Now the preparation of the Antimony is in this wise.

First take finely powdered Hungarian (which I consider the best) or in case such cannot be had other fused ♂, in weight 1 lb., drop it gradually into a red hot crucible, and let it fuse clear.

When this happens have ready a thin iron rod well red hot and stir it into the fused ♂ and it will cause an ebullition and eat into the red hot iron as much as is needful in the space of half of a quarter of an hour, that generally 1 lb. of ♂ takes as much as 14 to 18 & even 20 Loth or half ounces. When that is done throw a handful of well dried Sal Niter upon it, and as soon as the Niter is fused it must be poured into a casting cone, and you will find a solid Regulus, of about 18 to 20 or 24 Loth or half ounces in weight. This Regulus must be purified twice with Niter and Tartar and it will show a fine star, especially the Hungarian. Of Niter must be taken 1 lb. and of Tartar 8 Loth or half ounces and mixed together, therewith the martial regulus is purified, and reserved for the following use.

The other Reguli that are also needful for the high Work, that Mars add ⊖ed ♄ be brought, are to be made in this wise.

Take 1 lb. finely ground or pounded ♂ and mix thereto 1 lb. finely powdered niter and 3/4 lb. pounded Tartar, mix it together, and put it into a large mortar of cast iron in spoonfuls and with the first spoonful throw in a red hot cinder well alight, and let the mixture decrepidate. When it

130

has decrepitated, throw in some more, et cosi
successive che si calcina Sulfo, when it has cooled
pound the matter and let it fuse well in a red hot
crucible well covered that it may fuse quite clear,
when you will find a Regulus, weighing about 8 to 10
half ounces, which must once again be rectified with

⊖ & ♀. The scoriae remaining upon the Regulus
must be reduced to powder and placed on blotting
paper to filter, pour thereon common well-water, and

this so long, until some of the △ can be traced in
the common water, which can soon be discerned by

pouring in of wine vinegar. The △ filter & wash it
with common water. Accioche si levano totalmente le

salie sal △, quale in gustare, si puo cognoscere,

and in this wise you get a fine deep-red △ of ♂,
which you must Dry and keep for further use.

Those feces that remained on the paper, let them
again dry, powder them, and mix thereto equal weight
of finely powdered Tartar and let it fuse in a
strong heat, and you will get a very fine Regulus of
high lustre, which up to now has become revealed to
but few, & is considered an arcanum; now this
Regulus is the first that wholly unites itself with
Mars, although the first which was precipitated with

⊖ & ♀, and is got by the first fusion does also
perform its own part, yet nevertheless this one is

far higher esteemed, on account of its △ being
almost wholly fixed and its delicate mercurial
quality. The scoria remaining on this second

Regulus must again be pounded and put to filter, and you get again a ⚨ high in colour.

When no more ⚨ can be got, take the faeces, mix with some Tartar, and you will see whether it will still give a Regulus, and should it give no Regulus, then take the feces and put them in a wide crucible to calcinate, until they become quite ashen or whitish, then put them to filter with common boiling water, that the ☽ antimoni may be extracted. Let the water evaporate in a glass bowl, and you will find the ⊖ ♁ii, and if you would have it better and more clear, it is purified by frequent solution and coagulation, also with rectified S. V. or S. V. Tartarisati, but I for my part consider common well water best. This now is de ♁ how the ☿id est the ⚨ of the Regulus can begot.

But the Mercury for the high work of the universal particular Tincture is prepared in this wise, whereof also very clear description is given in the writings of John de Monte Snyder namely in his treatise de Metamorphosis Planetarum in Ch. 6, 22, 28, and other chapters can be seen and read the same kind viz.

Take 1 lb. ☿ washed clean with ⊹ & ⊖ common; grind it with 1 lb. good ♑, and 1 lb. common Salt, grind it all to powder, fiat pulvis, fiat ignis, fiat exaltatio as John de Monte Snyder mentions in Ch. 28 when all has been reduced to powder put in in a convenient subliming glass and

sublimate the ☿ at first with gentle fire until the steam & vapour is past, but afterwards with strong heat, that it may rise to the top quite white, and

in this wise must the ☿ be clarified three times, for thus also avers Monte Snyder in Ch. 27 when he says: Volatile sursum, bis vel fer.

Now when the ☿ has been on this wise prepared,

then take thereof 1 lb. finely powdered good ♁ and pour the same into a convenient retort, place it in a sand bath, and give it at first a gentle fire, and a few drops of phlegm will pass over, but afterwards the lac virginis, and when there is no more milk, some drops of blood will come over, then the fire must be increased, and you will get more blood and afterwards a few drops of mercury vive will come, then give it strong fire, from above as well as from below for 4 or 5 hours then goes the cinabaris in collo della retorta, then let it cool, keep the lac virginis separate, the mercury vive also separate and lastly also the cinnabar separate in a vessel. But in case the lac virginis should not be clear and

pure, then take Regulus ♁ Martialem 12 half ounces

let it fuse, and put thereto 4 fine ♀ plate, let fuse well together, and when it is well fused pour it out and pound it small, put it into a retort and pour thereon equal weight of lac virginis, distil it at first with gentle fire, and at last with strong heat, and the milk become somewhat yellow, which must be animated in such a manner:

133

Recipe de Cinabrio 3 or 4 half ounces, put thereto 1 half ounce ⚹ ♀, pour or place thereon 16 half ounces de Menstrui; let it dissolve at a gentle heat and the milk will become blood equestoe' Mercurio noster sedgsiato (?) ▽ el fons vitae, verus ☿ios noster essentificatus in which fountain the metallick ⊖ & ⚹ , are apt to be purified and joined together, which manner now however by means of the magical elements, that are the Keys to health and wealth, all metals are to be reduced into ⊖ & ⚹ or how the metallick soul is to be got, and also the fixed Sal metallica to be conquered, the process depends more on the manual work, but as the Science without the work and reciproce the work without the Science brings no fruits, such Science and Manipulation, as far as is possible, is herewith clearly described.

The magical elements or the threefold magical fire is ◑, ⚹ & Regulus ♂ which John de Monte Snyder calls ☿us Saturni, il ♀, si deve prendere accioche la anima, id est ⚹ metalica conserved before the fire.

John de Monte Snyder in Tractatis de Metamorphosis Planetarum[5] Ch. 15 thus describes the Magical Fire.

---

[5] The R.A.M.S. Library of Alchemy Volume 31. -PNW

With these words Vulcan prepared an artistic firework, which was made of an unkindled fire, a fiery air and a vegetable salt.

N. B. The unkindled fire is ♁, the fiery air is ◐, and the vegetable ⊖ is ♄, and is thus observed in weight.

◐   9 half ounces, - 3 thirds

♁   6 half ounces, - 2 thirds

♄   2 to 3 half ounces, - 1 third

Grind all together fine, let fulminate & keep.

The Conjunction and Amalgamation with the Metals and our ☿ id est regulo is done in this wise.

Regulus Martiale, ben purigato si prende 5 lotti et del ☉ 1 loth si fonde bene in sieme, poi si getta al cannlle e si tritura solilmense, e si serva per foliminare,

Regulo Martiale 4 lottoni & of the ☽ 1 loth sit amalgamo in igne.

Reg. ♂ 3 loth & de ♀ 1 loth, ♃ et ♃ .... 3 lotoni to 9 loth of the martiale Regulo. Regulo fatto per se, can ◐ et ♄ et primario il regulo ex scorijs, uti dictum est, take 4 loth powder such in mortano di fero, when this is done take thereto 2 loth thin lamels of iron, the thinner the better, lay them into a crucible with cover & let it get well red hot, then put in such powdered Regulus in spoonfuls & let it fuse well with great heat, but this sign must be observed that as soon as the

Regulus with the ♂ is in clear fusion then he has taken sufficient to himself, and it must then be cast, which will take place in about half an hour. Break the ingot and it must be white in colour, that is one sign. The other sign is this, I have taken 4

loth Regulus and 2 loth ♂, and the fused mass will weigh at least 5 1/2 loth, then I am assured that

the ♂ & Reg. are become one, which mass must be

reduced to powder and kept for Destruction. In this wise are reduced all metals in materiam primam id

est into our ☿ and is called the conjunction of the philosophical heaven with the terrestial planets.

Now how these conjoined metals are destroyed with the previous Described fulmine radicaliter in aperto Igne, the weight that must be observed, which is of great importance, the following is a clear report and actual experience as namely to the conjoined mass of gold and martial regulus of half ounce of these clear powdered mass are required of the aforedescribed fulmen.... 8 loth (or half ounces).

To one half ounce of the Lunar is required of the fulmen...6 loth.

To one loth of the Venereal .... 5 loth of the fulmen is required. Of the reg. el ♂ to one loth...6 loth fulmen; of the ♄ al ♃, to one loth of such conjoined mass is required of the fulmen 4 loth the same also if you would fulminate the Martial Regulus.

Now when the fulmen & the conjoined metal have been mixed together then it is needful to have at hand a crucible as red-hot as is possible surrounded with red-hot charcoals. In such red-hot crucible the aforesaid fulmen with the conjoined metals is thrown in by spoonfuls, but that at all times that which has been thrown in before must be in fusion. Now when the whole matter has been thrown in, then take great care that the fire does not diminish, the better the fire and the stronger it is, the more throughly will the metal be destroyed, but on the contrary if the fire is too small the matter remains & hardens to a black mass, which can hardly be any more brought into fusion except with the most violent heat and projection of Sal Niter, but which cannot be done without Danger. Therefore observe as a principal rule to keep a good heat in the destruction of metals.

Further let it be known that a certain sign must be observed that the metal does not burn away or be reduced to a black slag, for when the metal in strong heat becomes a vitrum then all is lost, for you get then neither 🜍 nor 🜔 but only a totally burnt matter, that cannot be used to universal or particular work and must be thrown away: To prevent this is of consequence to know the correct

manipulation and observe this sign namely when the
thoroughly mixed matter has been thrown into the
Crucible and is continually standing in fusion, you
will continually see for a certain time a flashing
smoke above the crucible and not be able well to see
the fusion of the matter, then is the time to cast
and you will find a blackish mass and lowest at the
bottom a little of the conjoined Regulus, the less
there is the better, then you are assured that

neither the ⚷ nor the ⊖ have been endangered in
the fire, as there was still some matter left for
the fulmen to lay hold of.

The fulminated matter must now be reduced to
powder, and placed to filter with cold common well
water, the water filtered through is put together
and some vinegar is mixed with it then the water
becomes and appears as blood, in this wise the

metallic ⚷ can be easily got by filtering the
common water and acetum. But the following must
thereby be observed, namely that all saline

admixtures of the ⚷ must diligently and well be
washed away with common water, which can not only be
known by smell and taste, but also the following
sign will be observed, namely, when the common water

has been poured on the filtered ⚷, and the water is

filtered through, pour a little ☩ into it, if the
colour of the water is green or yellowish then there

is still some saline admixed to the ⚷, but if the
water gains no other colour than ordinale mente,

then the 🜹 has been well cleansed from all
Salines.

Now when there is no more 🜹 in the fulminated
matter, then let the yellow feces dry, when they are
dry, rub to powder and fulminate it ad 🜔 with
following manner:

🜓   16 loth (half ounces)

🜹   10 loth (half ounces)

🜩   2 loth (half ounces)

This is altogether pulverized and put thereto 6
loth well desiccated id est well dried feces
mentioned above and mix, then throw into a red hot
crucible, and let it fuse in strong heat for 1/4 of
an hour or more, then pour it out, pulverize, and

put it to filter and with common water the little 🜹
with addition of aceti is washed away. Then dry the
feces, pulverize them and in a wide crucible put
them to calcinate, the crucible being covered, and
let it thus calcinate well for 12 to 16 hours,
namely until it becomes whitish or ashy colour, then
put it again to filter.

Filter the Salt well with hot water until there
is no more Salt therein, qual nel gustare delle feco
si puo conoscere, the water filtered through put
into glass basons on hot sand and let it gradually

139

evaporate until the ⊖ appears quite dry and white, then it is ready.

N. B. The Martial Salt will commonly fall somewhat reddish, except the first which give a quite white ⊖.

But if you would have these salts higher and more spiritualized they should be rectified with well rectified Spirit of Wine or else with rectified aceto or with common well-water, and their colour will be increased, but I hold it advisable if they appear clear and white at the first time keep those for although one may dissolve and coagulate such salts they will be found in same weight almost.

The feces remaining from the filtration of the Salts, are kept and when there is sufficient it should again be tried with the fulmine, whether it still gives off a ⊖, if not, it can be cast away.

In this described manner, all metallic and mineral salts are conquered and gained.

Now follows the Composition of the Universale Generalissimo.

In this wise: Take of the fixed and clear ⊖ metalico, grind it on a clean porphyry, thereof one half ounce, or as much as you will, upon one half ounce of ⊖ put 5 or more or 6 half ounces lacle Virgineo place it in a gentle heat for some 40 hours, or at the longest for 3 days and nights, and

the ⊖ therein will dissolve, as much as is
agreeable to the milk, then pour it off and keep it
in warmth until the following conjunction.

Similarly take two half ounces 🜍 and pour
thereon 10 to 12 half ounces of Milk, and let it
stand as long as the ⊖, and the 🜍 will become
opened in the milk, when this has happened, pour the
two solutions together into a long necked phial
close the mouth of the phile well with a cork and
white wax, and place it at first ad digestionem into
a very gentle heat for 30 or 36 days and nights, and
the matter will become quite black and thick like
pitch, then give it a stronger heat, that it begins
to come whitish or ashen grey, then augment again
the fire, that the matter gets quite white. Now when
it is become quite white (but before that many and
wonderful colours will appear) then give it strong
heat, till it becomes quite red. When it has been
brought to quite a red mass, and the matter on the
bottom of the phial remains quite fixed, then it
must be fermented and imbibed.

For example: I have taken of the solution of the
⊖ & 🜍 circiter 20 half ounces, and put thereto 1
upon 10, as namely 1 loth (half ounce) de 🜍 ♀ is.
and 1 loth de 🜍 ♂ , together with 2 loth lac
virgineo essentificato, and let it thus stand until
it again become quite red and fixed.

N. B. One to 10 is thus to be understood that
namely to 10 loth of tincture 1/2 loth de 🜍 ♀ is,

et ♂ aa. , with one loth ☿ essentificati must be taken.

To this massa of 24 loth when it has to be fermented for the second time is done as follows:

I take fine Pr. ♂̄, three times passed ☉ dissolve the same in ♈, precipitate it with Sp. ⊖ is, or Tartari, lixivate the gold calx well and throughly, that no acid remains therewith, powder it subtilly on porphyry, to this tincture of 24 loth you must have 2 loth gold calx, the 1 to 12 is the ferment. These 2 loth ☉ are mixed with 4 loth decocle Virgineo animiato, and put to the above described 24 loth Tincture and fixed together, now when it has again been fixed, then imbibe it again cum lacte animato with the fifth part of the weight of the tincture, namely, to 30 loth of the before described universal Medicine I take the 5th. part pro imbibitione de menstruu, namely 6 loth that is the first imbibition.

To the second   7 1/4 loth

To the third   8 1/2 loth

To the fourth   10 loth

Eat sic consequently each time with the fifth part of the already fixed tincture, that in this wise the Tincture in quantitate et qualitate in infinitum sau be augmented, and this Tincture on complete imbibitions tinges 1 part to 1000 parts of all imperfect Metals into ☉. And the oftener such

142

Tincture is dissolved and coagulated the higher it tinges. But the imbibitions must always be done with menstruo essentificato, which although it has been mentioned before, but there is a better and higher animation of the Menstruu made, if the red Tincture that already been got and is made in that wise.

Take ☿ ♀ 2 loth, of the Tincture 1/2 loth grind it well together and add 12 loth of the white milk, so that it may remain in gentle heat for two days and nights, then decant and use per imbibitiones ad infinitum.

In conclusion there is a still higher multiplication, which philosophers and adepts keep very concealed, namely they say by dissolving and congealating is the strength increased. I personally gained it by extra favour of our Adept.

Take of the fermented red tincture post of mam Imbibitionem 1 half ounce add thereto 10 half ounces menstruo animalo, let it dissolve in gentle heat, that it may go perfectly black and intoputrefaction, then encrease the fire until the perfect red, the longer with stronger fire which happens within two months, and one part tinges 10,000 parts, and this is the true Multiplication that always encreases ten-fold in mantitate et qualitate forever, almost beyond computation.

But as concerns human health this must be observed, that such de primo ordine before it has been fermented, id est, when it comes to the perfect red, of the same 2 to 3 and at most 4 grains de 2 ordine, when it has been 7 times imbibed one to two grains, de 3rd. ordine so such tenfold in quantitale of qualitate encreased that a mustard seed therefore is sufficient. Take it in any convenient liquor or cordial. Come il adepto misustru isse.

Now follows the other Universal Work.

Which is not generalissimo, but only such a universal work that changes all imperfect metals into constant gold and silver, but not in such form and fineness as the before described, but it can be augmented in qualitate et quantitate per imbibitiones, that one part of such tincture can transmute several hundred parts of imperfect metals, and the conjunction and fixation is done in the following manner:

Take lunar ⊖ 1 loth, Sulphur ♀ 3 drachmas, Sulphur ♂ 3 drachm. Grind it together to finest powder, put it in a phiel con il collo longo, and pour in 6 loth of the milk animated with Antimonial Cinnabar.

Close the glass well with a cork and with white wax, place it at first for some 20 days into gentle heat, and it will quite at the beginning enter into putrefaction, and thus in gentle heat dissolve and coagulate, when you observe that the black is about to pass away, then you must give another grade of fire, until it begins to have divers colours, then give stronger heat, namely the third grade, and you will get the perfect white, when the white has taken place then give the fourth grade of the fire that you may see the red. This red has no real stability, for it will now turn whitish now reddish then is time that it must be fermented, in such a

144

manner when there is of the Tincture 8 ½ loth, 1
take to it 1/2 loth 🜍 ♀ and 1/2 loth 🜍 ♂
with 2 loth of the animated milk, put it in
digestion for some days and nights, at first in
gentle afterwards in stronger heat, namely in the
fourth grade, and get the perfect fixed red.

This is fermented to become Tincture in the
following way:

Take ☉ reduced to calx by ℞ and lixivated of
all acidity, if the Tincture is in weight 11 1/2
loth, I take of the ☉ calx 1 loth, and 3 loth of
the menstruo essentificato, put it together and when
it has come to coagulation, then I have 15 1/2 loth,
then I take again the other imbibition as with the
5th. part, that is, 3 loth of animated milk, let it
again together coagulate and fix.

In the third imbibition take	3 1/2 loth
In the fourth imbibition take	4 1/2 loth
In the fifth imbibition take	5 loth
In the sixth & seventh	6 loth.

After termination of all imbibitions take out
the tincture, divide the same into other and larger
glasses, and augment the same again with the fifth
part of the fixed tincture in quantitate et
qualitate ad libitum. This Tincture after seven
imbibitions tinges 1 part to 100 parts of all
imperfect metals into constant gold, and the oftener
it is imbibed and coagulated, the higher it gets in

its operation, that one part transmutes several hundred parts of impure metals into fine gold.

Then it is increased in quantitite in the following manner:

If you put of such tincture 3 parts to one part of pure gold in a crucible that well endureth the fire, lute the same well and set it in a good wind and melting furnace and keep it in fusion for three days and nights, when all will become a vitrum, of which 1 part thrown on 100 parts of heated Mercury ex destillatione of the white milk will still become tincture, and tinges 1 part also some hundred parts of other Metals into ☉, and this is an increase in short time in quantitate, but the Mercury must be made rather hot before you project the tincture upon it.

But per corpo humano this universal medicine is used against all diseases, unfermented, from 4, 5 or 6 grains. Post eni Imbibitiones 1, 2, or at most 3 grains, but if it has been frequently augmented you give only one grain.

Particulare ex universali.

Take ◯ ☽ or Sal ♂ 1 loth, ♄ ♀ 2 loth mix well together, with 1/2 loth well elixivated silver calx, grind it well together, and add thereto 3 1/2 loth of milk animated with ♄ ♀ et Cinnabrio ♁i in this wise you put first the milk at the bottom of the phial, which is best if it have a flat bottom, and thereon put the mixture de ⊖ ♄ et

146

fermento; Close the phial with a cork and white wax, and put it at first 2 or 3 weeks in gentle digestion that through putrefaction it may get to perfect coagulation and ash colour then increase the fire until the matter gets quite white, when it is quite white then give it another imbibition with tacle animato, that it may become more easily fusible than with 2 loth of the milk, or also a little more, when it has come through putrefactio ad albedinem, it is

a medicine per ☿ comune ad augmentum perpetuum in this wise that you take of this Medicine 3 parts and

put one part of ☿ com. that has been washed with ⵚ

and common ⊖, grind it together and put it ad digestionem for 14 days or 21 days at most, and the

☿ will become perfectly fixed and able to pass cupellation, in such wise that you can set such augmentum ad libitum and the augmentum perpetuum is

always continued with a third part of ☿ com.

But there is still a better augmentation that

tinges all metals into ☉ to be done with this particular process namely; when it has arrived at the perfect white you can at once imbibe and ferment

it with both 🜍 de ☿ or ♂ or also with both together and 2 loth animated Menstruo that has melted with the fourth grade of the fire to perfect red, and it is also very good and very serviceable when it has got to the red, that it should again be

imbibed with 1 loth 🜍 ☿ and 2 loth de Menstruo and again be brought per omnes colores ad rubedinem, then it tinges one part to 10 parts of imperfect metals except ♂ into constant gold, and such imbibition only with addition of the 5th or 4th. part of de Solo Menstruo essentificito; can such tincture be augmented so high in quantitate et qualitate, that 1 part will tinge several 100 parts.

The other Particular.

Take well elixivated calx of Luna 4 loth 🜍 ♀ et ♂ ana 1/2 loth, De Menstruo essentificito 2 1/4 loth put it together let it digest for a time of 4 weeks at first gently, then with strong heat, and all the ☽ a will become good ☉.

The third Particular.

Take ☉ calx 2 loth, 🜍 & ⊖ de Marte 1/2 loth, ☿ 1 1/2 loth, Menstruo essentificato con Cinnabrio 1 1/2 loth, put it together for a month, at first gentle and then strong heat till it gets red, then imbibe it well with 1 1/4 loth id est 5 Drachmas of milk and by strong fire let it get red at once. Of this medicine 2 loth projected upon 2 mark of ☽ standing un fusion gives 10 to 12 loth ☉ and when this medicine has been frequently imbibed with

Menstruo animato, it tinges also ☿ & ♃ into constant ☉.

Now follows how the true Aurum Potabile also potabile can be made.

Firstly make a pure ⚧ de ☉, or ☽, dissolve the same in Menstruo nostro animiato con Cinnabaro, leave it for four days and nights in quite gentle heat, and the Menstruum will dissolve the most noble ○lar or ☽ or ⚧ and the menstruum will become cloudly, upon the same pour rectified Spirit Vini Tartarisati, leave it to putrify in Balneo for a month, till the S. V. becomes highly coloured, distil the S. V. off with gentle fire, and you will get the true Aurum Potabile, which cannot again be reduced to a corpus; Of this use 1, 2, or at the most 3 Drops. And in this wise all Metallick △⚧ as well de ♀ as of ♂ can be brought to potability and used as high Medicine.

Further follows how all gems, as diamonds, rubies, and emeralds and sapphires can be augmented, tinged and brought into a true Tincture.

Take of the second or third Regulo ♂ made of the feces, when the ♂ is fused with ◐ & ⚥ and the ⚧ has been washed off, as is clearly described on page 6, take the same regulo, powder it, then take thereof 3 or 4 loth and of the gem, be it Diamond, ruby or what you will, mix it well together and put it in a covered crucible should have a hole in the centre, place it in a fire of calcination for 24 hours, then open the Crucible and feel with an iron rod if there is any regulus left therein, and in

149

case the regulus therein is in fusion then let it go
off with a good fire and stirring continually, till
all is reduced to powder, then take it out, let it
get cold, and put it on an iron patola, let it get
red hot in the fire, and cast it while red hot into
the hereafter described water, and the gems will
become perfectly calcined and can be reduced to fine
powder by grinding, and the more frequently this is
done of getting it red hot and then quenching it the
better it will be.

Common distilled well-water by weight 4 lbs.
therein dissolve ✳ 8 loth, Sal Commune 8 loth, and
Sal Nitro 8 loth.

In this water is done the quenching of all kinds
of gems.

Although there are also other calcinations of
gems according to the ways of some adepts as with

pure and clear ♇ ☿ with addition of a fixed Sal
Commune, but these are long and difficult processes
and the before described way is the best because
given by experience and proved to be real by adepts.

Now when the gems have been reduced to subtile
powder, take of the same one part as for instance:

Gem powder	1 loth
◐	5 loth
♇	3 loth
♀	1 loth

this all well mixed together and 1/2 loth Cinnabrio

☿ ground into it that the gems may be more
throughly attacked, and such composition is thrown
in spoonfuls in a very red hot crucible and with
best heat as possible kept in continual fusion for a
quarter of an hour, then pour it out on a grinding
stone reduce it to powder, and put it to filter with
common well-water that all acidity may go away, but

the sulphur you must conquer by pouring ✠ into the
first water running through the filter, of which by
weight very little will be gained.

     The feces are thus calcined ad salom but they
must at first be well desiccated, else they are not
good for fulmination:

     Take the feces   4 loth

        ⊖   8 loth

        ⚴   4 loth

        ♄   ½ loth

mix it well together let it well fuse together for a
time that one could say several times the Lords
Prayer, then pour it out and with diligence reduce
it to fine powder, then per filtrum deprive it of
the saline which can be discerned by tasting.

     When this has been done, take the faces let them
again be well dried and put them to a good calcining
fire in a wide crucible, and calcine for 24 hours,

151

that the crucible may stand at all times in a good heat, then take it out let it cool, pulverise, and put it to filter with boiling water. Put the water into evaporating basons till the ⊖ is well dried, in this wise you get the ⊖ & ⊕ of all gems.

The lac virginis is animated therewith in this way: Take of the white milk which is quite pure which must not be animated with ought else than the ⊕ of gems, as follows: 1 loth of the ⊕ of gems is taken to 8 loth of the milk, place it into gentle heat, that it may dissolve then let it stand in a closed vessel for three days and nights, and the menstruum will absorb the most noble ⊕ and essencified therewith.

But the Conjunction of the Tincture is done in the following way:

Of the white unfermented earth, generated by the conjunction

of ○ ☽ et ⊕ ♀ cum menstruo nostro and called abbificatio;

take abbificatio   4 loth

of the gem ⊖   1/2 loth

the gem ⊕   1/2 loth

Of the rightly animated Medicine   2 loth.

This all well mixed, namely, la terra alba ⊖ et ♧ together put into a phial and then the menstruum poured thereon, set it at first in gentle heat for 8 days and nights, then continue with stronger fire to the highest white and it will tinge Crystal, made soft or red hot, to the highest Diamond, in the red colour should it have been fermented with Diamond, to true Carbuncle, in the green colour if it is so fermented with emerald, to the highest green Emerald.

In the blue colour to Sapphires and lastly should it be fermented with the ⊖ et ♧ delli rubini and brought into these, it will also tinge Crystal to such high Rubies as in lustre will not be inferiour to diamond.

This tincture is always animated with its like animated menstruo continually multiplied and imbibed. Always if there is 6 loth of Tincture in weight like 1 1/2 loth of the animated milk.

One more Particular pro Corpo homano to be used in all Diseases,

its dose is from 1, 2, to 3 grains.

Namely Sal ♂   2 loth

♧ ♂   1/2 loth

♧ ♀   1 loth

♧ ♃   1 loth

153

This all well mixed together put it in a phial,
and pour thereon 1 1/2 loth de menstruo animato, and
digest it within two months time at first in gentle
and afterwards in strong heat to redness, it is an
excellent medicine as before mentioned.

It is augmented con ⊖ ♃ et ♁ ♀ in this
wise:

Of the Tincture there is 7 loth, take thereto
⊖ ♃ & ♁ ♀ -- 1/2 loth, menstruo animato 1 loth
and in this way it is always augmented.

Merton 24 IV. 1901   J. K.

A Word from the Publisher

Thank you for purchasing this small work from The R.A.M.S. Library of Alchemy. During his lifetime, Hans Nintzel was dedicated to the identification, acquisition, study, retyping and, when necessary, translation of what he considered to be the most important known works on Alchemy. Hans was assisted by his sparse network of fellow Alchemists, all members of the Restorers of Alchemical Manuscripts Society (R.A.M.S.). I was an active member of R.A.M.S.

My goal is to publish all of the works originally made available through R.A.M.S. as photocopies. To facilitate this, I have chosen to have the books professionally printed. I also have a few titles that I intend to add to the original R.A.M.S. Library, selected by strict criteria established by Hans.

The works from the original R.A.M.S. Library are republished by R.A.M.S. Publishing Company in the collection, "The R.A.M.S. Library of Alchemy," with permission of the Estate of Hans W. Nintzel.

If you have a work on Alchemy that you believe should be a part of the R.A.M.S. Library, please contact me through R.A.M.S. Publishing Company.

Philip N. Wheeler

www.ingramcontent.com/pod-product-compliance
Lightning Source LLC
Chambersburg PA
CBHW080812180526
45168CB00006B/2411